T0290958

Multiscale Geographically Weighted Regression

Multiscale geographically weighted regression (MGWR) is an important method that is used across many disciplines for exploring spatial heterogeneity and modeling local spatial processes. This book introduces the concepts behind local spatial modeling and explains how to model heterogeneous spatial processes within a regression framework. It starts with the basic ideas and fundamentals of local spatial modeling followed by a detailed discussion of scale issues and statistical inference related to MGWR. A comprehensive guide to free, user-friendly, software for MGWR is provided, as well as an example of the application of MGWR to understand voting behavior in the 2020 US Presidential election. **Multiscale Geographically Weighted Regression: Theory and Practice** is the definitive guide to local regression modeling and the analysis of spatially varying processes, a very cutting-edge, hands-on, and innovative resource.

Features:

- Provides a balance between conceptual and technical introduction to local models
- Explains state-of-the-art spatial analysis technique for multiscale regression modeling
- Describes best practices and provides a detailed walkthrough of freely available software through examples and comparisons with other common spatial data modeling techniques
- Includes a detailed case study to demonstrate methods and software
- Takes a new and exciting angle on local spatial modeling using MGWR, an innovation to the previous local modeling 'bible' GWR

The book is ideal for senior undergraduate and graduate students in advanced spatial analysis and GIS courses taught in any spatial science discipline as well as for researchers, academics, and professionals who want to understand how location can affect human behavior through local regression modeling.

Multiscale Geographically Weighted Regression

Theory and Practice

A. Stewart Fotheringham, Taylor M. Oshan, and Ziqi Li

CRC Press is an imprint of the
Taylor & Francis Group, an informa business

Designed cover image: © A. Stewart Fotheringham, Taylor M. Oshan, and Ziqi Li

First edition published 2024
by CRC Press
2385 Executive Center Drive, Suite 320, Boca Raton, FL 33431

and by CRC Press
4 Park Square, Milton Park, Abingdon, Oxon, OX14 4RN

CRC Press is an imprint of Taylor & Francis Group, LLC

ISBN: 978-1-032-56422-7 (hbk)
ISBN: 978-1-032-56423-4 (pbk)
ISBN: 978-1-003-43546-4 (ebk)

DOI: 10.1201/9781003435464

Typeset in Times New Roman
by Apex, CoVantage, LLC

Access the Support Material: https://sgsup.asu.edu/sparc/multiscale-gwr

ASF: Elizabeth Helen Fotheringham

TMO: Ramona Belfiore-Oshan

ZL: Li-Ling Chang

Contents

Acknowledgments

The authors would like to acknowledge the assistance of Dr. Mehak Sachdeva in preparing some of the figures in the book and also for reading preliminary drafts of some of the chapters. We would also like to thank the Human, Environment, and Geographical Sciences (HEGS) section of the United States National Science Foundation for their generous support of much of the research that is contained within this book. The support of grants (1758786, 2117455) is gratefully acknowledged.

Authors

A. Stewart Fotheringham is Regents' Professor of Computational Spatial Science in the School of Geographical Sciences and Urban Planning at Arizona State University. He is also director of the Spatial Analysis Research Center and a Distinguished Scientist in the Institute for Global Futures. He is a member of the US National Academy of Sciences and Academica Europaea and a Fellow of both the UK's Academy of Social Sciences and the Association of American Geographers. He has been awarded over $15 million in funding, published 12 books and over 250 research publications. He has over 38,000 citations according to Google Scholar as of July 2023 and is one of the top-cited academics in the field of geography. He has been awarded the Lifetime Achievement Award by the Chinese Professional Association of GIS and the Distinguished Research Honors Award by the American Association of Geographers.

Taylor M. Oshan is assistant professor in the Center for Geospatial Information Science in the Department of Geographical Sciences, University of Maryland, as well as an affiliate of the Social Data Science Center, the Maryland Population Research Center, and the Maryland Transportation Institute. His research focuses on developing and applying multiscale methods and local statistical models, particularly of human processes within urban environments, to understand how relationships change across different spatial contexts. He also leads projects to develop open-source tools for spatial analysis, including the core algorithms for the multiscale geographically weighted regression software among others. He has published over 25 peer-reviewed manuscripts and collaborated or led funded projects totaling over $2.3 million. He was elected as a board member in 2021 for the spatial analysis and modeling specialty group of the American Association of Geographers and joined Applied Spatial Analysis and Policy as a co–editor-in-chief in 2023.

Ziqi Li is assistant professor of Quantitative Geography in the Department of Geography at Florida State University. His research focuses on the methodological development of spatially explicit and explainable statistical and machine learning models, and he is one of the primary contributors to the field of multiscale geographically weighted regression. He has published over 20 peer-reviewed journal articles in these areas. He is a winner of multiple prestigious international awards, including the Nystrom Award by the American Association of Geographers (AAG) in 2021 and the John Odland Award by the Spatial Analysis and Modeling Group of AAG in 2020.

Preface

It is 20 years since the publication of the seminal text on geographically weighted regression (GWR) by Fotheringham et al. (2002), almost 30 years since the first crude articulations of this approach appeared (Fotheringham & Rogerson, 1993; Rogerson & Fotheringham, 1994), 40 years since one of the authors wrote about spatially varying distance-decay parameters and parameter sensitivity to multicollinearity (Fotheringham, 1981, 1982), and 50 years since the publication of Casetti's article on the expansion method, which set the ball rolling in terms of thinking about spatially varying relationships (Casetti, 1972). During this time, a great many advances in regression-based, spatially local, statistical modeling have taken place, new frameworks for local modeling have appeared, a huge number of empirical applications have been published, and advances in computational methods have made the calibration of complex spatial models on reasonably large datasets feasible. It thus seems timely to capture, assess, and explain these developments in one publication focused on a recent major advance in this field, that of multiscale geographically weighted regression (MGWR).

Although there is a focus on statistical methodology in this book, perhaps its greater contribution is elucidating in a nontechnical way the concepts behind what is termed 'local modeling'. Why might some relationships vary over space? What could cause such variation? What are the ramifications of such variation? Only when one comes to accept that there might be a problem with a certain course of action does an alternative begin to have appeal. Consequently, a great deal of effort is made to explain the raison d'être behind local modeling, the concepts behind the particular local model form that is the focus of this book, MGWR, and the implications for spatial analysis and society in recognizing that a "*one size fits all*" mentality may not be optimal when it comes to modeling human spatial behavior.

The book covers a wide range of topics related to local spatial modeling in a regression framework. Chapter 1 begins with a discussion of what is meant by 'local' and why 'local' influences might be important in understanding human spatial behavior. This inevitably concerns another loaded term, 'context', which we discuss as a summary of various ways location might affect behavior, independently of any factors pertaining to an individual or a location that we can measure.

Chapter 2 covers the basic statistical principles behind MGWR including the concepts of 'data borrowing', 'data weighting', and bandwidths. It begins with a description of the precursor to MGWR, GWR, and then outlines how the model and its calibration differ when the restriction of a single, average bandwidth in GWR is dropped to allow covariate-specific bandwidths in MGWR. The chapter contains a summary of model diagnostics including those for model fit, hypothesis testing, and detecting potential problems.

Local models are '*big*' models in that they produce large amounts of output in the form of localized parameter estimates and associated statistics. Consequently, inference is a very important part of any empirical application of local models in order to separate signal from noise. In Chapter 3 we cover three types of inference that are essential in MGWR: (i) inference about individual, local parameter estimates; (ii) inference about surfaces of local parameter estimates; and (iii) inference about bandwidth parameters. Inference about parameter estimates from local models has to deal with both the multiple hypothesis problem and the local estimates exhibiting varying degrees of spatial dependency. We demonstrate how both of these issues can be addressed by the use of simple adjustments to classic inference. In terms of bandwidth inference, we show how Akaike weights can be used to compute confidence intervals around covariate-specific bandwidths.

Chapter 4 is concerned with the issue of process scale. First, it demonstrates how the bandwidth parameter in MGWR is an indicator of the geographic scale over which a process varies, and second, it reexamines the classic modifiable areal unit problem (MAUP) and Simpson's paradox through the lens of local modeling. Here, a focus on the properties of processes, in particular their degree of spatial heterogeneity, leads to new insights into the causes of both the MAUP and a spatial variant of Simpson's paradox.

Chapter 5 contains a demonstration of MGWR calibration using freely available and user-friendly software, MGWR 2.2. This is a very simple-to-use, yet very powerful, package for calibrating models by MGWR. The chapter also describes some of the computational issues behind the software and briefly describes other software available for calibrating both GWR and MGWR models.

Because it is very easy to be misled into interpreting essentially noise in local modeling, Chapter 6 contains an important series of warnings regarding local modeling. These are divided into pre-calibration caveats, calibration caveats, and post-calibration caveats and are designed to reduce the chances of producing empirical results that are difficult to defend. The chapter concludes by providing a seven-point checklist for any empirical application of MGWR (and indeed, any type of local modeling).

Chapter 7 contains an application of MGWR to voting behavior in the 2020 US Presidential election. This not only serves as a demonstration of the power of local modeling but also is a blueprint for any empirical application of MGWR by providing a checklist of actions that should be undertaken to maximize the robustness of the model outputs and interpretations.

Chapter 8 demonstrates the relationships between three modeling frameworks that can be used to assess the role of context on behavior: MGWR, spatial error models, and multilevel models. We demonstrate that the former has important advantages in terms of how it can be used to quantify contextual effects, and we also show that spatial error models and multilevel models can be considered as special cases of MGWR. This is important because not only do local models, such as MGWR, allow spatially heterogeneous processes to be measured, but they also produce residuals with relatively little spatial dependence. We also discuss how machine learning can be used to model locally varying processes and compare the performance of one type of machine learning model with MGWR.

To date, empirical applications of MGWR are almost exclusively restricted to continuous data and the assumption of Gaussian random errors. Chapter 9 describes several extensions to the MGWR framework including the setting of the MGWR model within a general linear modeling framework (Multiscale Geographically Weighted Generalized Linear Modeling [MGWGLM]) and an extension to include temporal weighting as well as spatial weighting (Multiscale Geographically and Temporally Weighted Regression [MGTWR]). Another extension is that of MGWR models in which the optimized bandwidths are specific to locations as well as covariates. Chapter 9 also contains discussions of the use of local models for prediction and also on the impacts of local models to the debates surrounding the role of replication and reproducibility in social sciences and also to the long-term schism in geography between nomothetic and idiographic approaches, which local models neatly bridge. The chapter also contains a summary of recent research investigating long-term problems of scale, such as the MAUP and Simpson's paradox, through the lens of local modeling.

Note

There is a website for support materials and additional resources, hosted by the authors—https://sgsup.asu.edu/sparc/multiscale-gwr

References

Casetti, E. (1972). Generating models by the expansion method: Applications to geographical research. *Geographical Analysis*, *4*(1), 81–91.

Fotheringham, A. S. (1981). Spatial structure and distance-decay parameters. *Annals of the Association of American Geographers*, *71*(3), 425–436.

Fotheringham, A. S. (1982). Multicollinearity and parameter estimates in a linear model. *Geographical Analysis*, *14*(1), 64–71.

Fotheringham, A. S., Brunsdon, C., & Charlton, M. E. (2002). *Geographically weighted regression: The analysis of spatially varying relationships*. Chichester: Wiley.

Fotheringham, A. S., & Rogerson, P. A. (1993). GIS and spatial analytical problems. *International Journal of Geographical Information Systems*, *7*(1), 3–19.

Rogerson, P. A., & Fotheringham, A. S. (1994). GIS and spatial analysis: Introduction and overview. In A. S. Fotheringham & P. A. Rogerson (Eds.), *Spatial analysis and GIS* (pp. 1–10). London: Taylor & Francis Group.

1

Introduction to Local Modeling

1.1 Setting the Scene

Research in many fields is prompted by the empirical observation that the values of most attributes vary over space and/or time. The earliest astronomers were guided by observing the night sky and noting the changes in the positions of certain stars or by observing changes in the direction of the rising sun throughout the year. Geographers have long been prompted to ask why certain physical features appear in some locations but not in others or why the distributions of almost all social characteristics of populations are not constant over space but exhibit variation, sometimes extreme. Epidemiologists study why the incidence of a disease exhibits both spatial and temporal variability and political scientists investigate the causes of variations in the support for one political party over another. The common thread across these research areas is a desire to discover why measured values exhibit variation—it is variation not stability that prompts us to ask "why?". For example, imagine that you wanted to know something about the climate of the United States but were given only the average annual temperature and rainfall across the country. These values would provide you with little information compared to the records on temperature and rainfall available for weather stations across the country and from which the US averages were generated. Variations in these local values would prompt you to ask questions such as "what causes rainfall to be lower in the southwest than in the northwest?" and "why is there a large range of temperature values throughout the year in the mid-west but not in Hawaii?" This seems a trivial and rather ridiculous scenario—why would anyone only report average values of temperature and rainfall and omit all the interesting variation behind these averages? However, this is exactly what has been happening for decades when spatial data are modeled. Traditional models of processes are global—meaning they report one parameter estimate for each relationship in the model. This is fine *if* the processes (relationships) being modeled are stationary—that is, they are the same everywhere. An increasing body of evidence suggests, however, this may not be the case, particularly when human behavior is being modeled. When processes vary over space, local models are needed to avoid reporting rather useless, and potentially highly misleading, average parameter estimates. This book describes the concepts behind, and the operation of, one type of local model—multiscale geographically weighted regression (MGWR).

DOI: 10.1201/9781003435464-1

The above paragraph highlights an important distinction between research focused on *data* and research focused on *processes*. Throughout most of its long history, human geography, for example, has been primarily concerned with data. Initially the focus was on mapping data but during the 1960s, statistical methods were introduced, and the analysis of spatial data came to the fore, exemplified by measures of spatial dependence and point pattern analysis. The idea that we could quantify issues such as the degree to which similar data values clustered was appealing and stimulated a huge literature (Moran, 1948, 1950; Whittle, 1954; Matheron, 1963; Paelinck & Klaassen, 1979; Cliff & Ord, 1981; Hubert et al., 1981; Anselin, 1988, 1995; Getis, 1991; Getis & Ord, 1992; Ord & Getis, 1995, 2001; Griffith, 2003). Around this time, human geography developed its first, and possibly only, 'law', which again focused on data and which is paraphrased as "*everything is related to everything else, but near things are more related than distant things*" (Tobler, 1970). Although this is not really a 'law' but an empirical observation, it stands as a description of a regularity that has achieved the status of a law within geography and succinctly captures the concept of data spatial dependency.

As geography's 'quantitative revolution' proceeded and spatial analysis became both commonplace and more mature, the traditional emphasis of the discipline on data gradually gave way to a competing interest in processes. It became insufficient to describe *how* some variable was distributed over space and the emphasis shifted toward providing an explanation of *why* a variable was distributed in a certain way over space. This '*why*' question led to an interest in the process or processes causing the observed distribution of data. Although this refocus on process in quantitative human geography now has a reasonably long history, nicely encapsulated by Hay and Johnston (1983), one problem immediately became clear. As 'process' is often interpreted as a sequence of events leading to a particular outcome, studies in cross-sectional spatial data lack the temporal dimension implied in this definition of process. Often, spatial data related to human behavior are observed at multiple locations but in only one time period. In contradistinction, studies in physical geography and economics often have the luxury of datasets that have a rich temporal dimension, although sometimes with limited spatial resolution. Under the assumption that processes are spatially stationary (the usual scenario, for example, in the physical sciences), the lack of spatial resolution is not a hindrance to the identification of processes if sufficient time-series data are available at even just a single location. However, what if the processes being studied are *not* spatially stationary (a likely scenario, as we shall see, in the social sciences)? In this situation, can we then replace the variations we typically measure in the temporal dimension with variations we observe in the spatial dimension to infer something about such processes?

As a consequence of the growing interest in process rather than form, there came a focus on the development of models, particularly explicitly spatial models such as spatial regression models (Anselin, 1988, 2002, 2009; Griffith & Csillag, 1993; Haining, 1993, 2003; Tiefelsdorf, 2000; Banerjee, 2003; Gelfand et al., 2003; Waller & Gotway, 2004; LeSage & Pace, 2010) and spatial interaction models (Olsson, 1970; Wilson, 1971; Fotheringham, 1983a, 1983b, 1984; Fotheringham & O'Kelly, 1989). Although spatial models have several purposes, undoubtedly the main one is to try to uncover what factors affect the spatial distribution of a particular variable or, rather crudely, to answer the question: "*Why are some values high and why are some low?*" This is typically done by making inferences about how covariates influence the dependent variable through the estimates of the parameters (or coefficients) obtained in the calibration of a model.

The important distinction between *data-focused* research and *process-focus*ed research focus is illustrated in Figure 1.1. In whatever subject matter we study, we understand that there are processes, which here we think of as being conditioned relationships, that lead to a set of outcomes that we measure. In the physical sciences, such processes can often be deduced from first principles, but in the social sciences, these processes are often unknown and have to be inferred from the measured outcomes via statistical associations. This latter situation is not ideal because of the problem of equifinality—the same outcome can result from different processes—but it underlies the complexity of dealing with the actions of human beings where universal laws of behavior are rare.[1] However, it can be argued that making inferences about properties of processes, and occasionally being wrong, is better than ignoring them. In this book, we focus on a situation with even greater complexity—not only do we have the problem of identifying the generally unknown processes that affect human behavior, but we examine the situation where these processes may not be the same everywhere. This again highlights the difference between modeling the physical environment and the human envixronment.

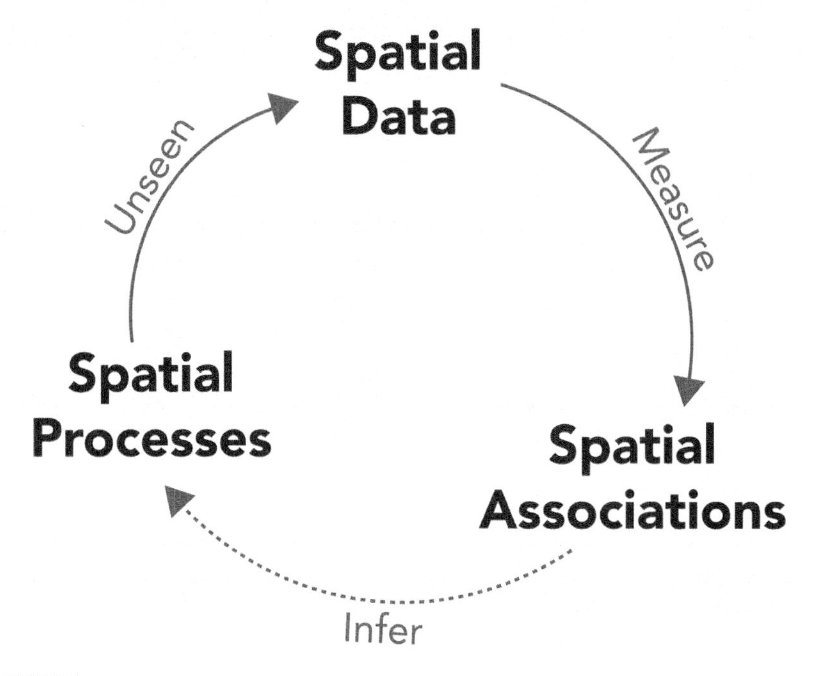

FIGURE 1.1
The Relationship Between Spatial Processes and Spatial Data.

1.2 Local Versus Global Models

From the origins of the quantitative 'turn' across many social sciences came a focus on relationships between attributes with regression-based models, as exemplified by equation (1.1), being especially popular:

$$y_i = \beta_0 + \beta_1 x_{1i} + \beta_2 x_{2i} + \ldots + \beta_k x_{ki} + \varepsilon_i \tag{1.1}$$

where y_i is the variable of interest measured at location i, $x_{1i}, x_{2i}, \ldots, x_{ki}$ are covariates, again measured at location i, β_0 is the intercept, $\beta_1, \beta_2, \ldots, \beta_k$ are slope parameters, and ε_i is a random error term. Each of the slope parameters represents the conditional effect of a change in the respective covariate on y and hence is an indicator of a specific process operating to contribute to the value of y observed at each location. Consequently, it is from the estimates of these parameters obtained in the calibration of the model that we make inferences about each of the processes that together create the observed distribution of y.

It quickly became apparent that models of the format of equation (1.1) were often inadequate when applied to spatial data because the distribution of the estimates of epsilon generally exhibited significant positive spatial dependency, violating the requirement of Gaussian regression that the residuals be identically and independently distributed. This prompted the development of various forms of what became known as *spatial regression models* exemplified by that of a spatial error model shown in equation (1.2) (Anselin, 1988; Gibbons & Overman, 2012; Kelejian & Prucha, 1998, 1999; Lesage, 2016; LeSage & Pace, 2010):

$$y_i = \beta_0 + \beta_1 x_{1i} + \beta_2 x_{2i} + \ldots + \beta_k x_{ki} + \lambda(\boldsymbol{W})\varepsilon_i + \mu_i \tag{1.2}$$

where ε_i is a spatially autocorrelated error term, λ is a parameter estimated to measure the spatial autocorrelation in the error terms ε_i, $\boldsymbol{W}$ is a spatial weights matrix, and μ_i is an independent and identically distributed (i.i.d.) error term.

A fundamental assumption in ordinary least squares or spatial regression models (and in almost any other form of model used prior to 2000) is that the processes being inferred through the parameters of the model are stationary over space. That is, we typically collect data from various spatial locations and use all these data to calibrate a model of the form of either equation (1.1) or (1.2) to produce a single estimate of each parameter. The implicit assumption is therefore that the process represented by a single parameter in the model is stationary over space. We term models that incorporate such an assumption *global*.

The widespread assumption that processes are stationary over space and that global models are appropriate for modeling human behavior was presumably a holdover from borrowing quantitative techniques, which had their origins in the natural sciences where processes are typically stationary over space. However, processes involving the beliefs, preferences, and actions of human beings may well vary according to location. Indeed, a huge literature exists supporting this idea (*inter alia*, Chetty & Hendren, 2018; Sampson, 2019; Darmofal, 2008; Plaut et al., 2002; Escobar, 2001; Diez Roux, 1998, 2001). To accommodate possible spatial process heterogeneity, various modeling paradigms have been developed by geographers and statisticians that overcome the limitation of global models by allowing the parameters in a model to vary over space, as typified by equation (1.3):

$$y_i = \beta_{0i} + \beta_{1i}x_{1i} + \beta_{2i}x_{2i} + \ldots + \beta_{ki}x_{ki} + \varepsilon_i \tag{1.3}$$

where x_{ki} is an observation of the kth explanatory variable at location i, β_{ki} is the kth parameter estimate, which is now specific to location i, and ε_i is a random error term. In this representation of the world, spatial process variation is accommodated by the flexibility of allowing each parameter to vary over space. Such models are referred to here as *local.*

Local models of spatial variation in the processes affecting spatial patterns can be divided into two categories: (i) models that require some *a priori* knowledge of the spatial variation in the processes being examined and (ii) models that do not have such a requirement and instead produce information on the nature of the spatial extent of any variation in processes from the data. The former typically require a number of spatially defined subsets of the available data to be specified by the analyst, and parameter estimates for each geographic relationship (i.e., process) are obtained specific to each subset of the data. Examples of models that fall into this category are multilevel models and spatial regime models (Anselin, 1988; Duncan et al., 1998). While the nature of the variation in parameter estimates from these regional models may provide clues toward the scale of geographic relationships, they do not provide explicit measures of process scale. Instead, they require scale to be specified *a priori.* They also assume that the relationships being modeled are constant within each predefined subset of the data but vary across the subsets, producing a discontinuity at the borders of units (Lloyd, 2007).

The latter type of local models allows for continuous spatial variation in the processes being modeled and estimates the nature of process heterogeneity directly from the data without the need to pre-specify clusters of data. This allows for a more flexible analysis of any spatial heterogeneity in processes. Examples of such local modeling frameworks include eigenvector spatial filter-based local regression (SFLR/ESF) (Griffith, 2008; Murakami et al., 2017, 2019), multiscale geographically weighted regression (MGWR, building on the earlier GWR framework (Fotheringham et al., 1996; Brunsdon et al., 1996; Fotheringham et al., 2002, 2017), Bayesian spatially varying coefficient (SVC) models (Gelfand et al., 2003; Banerjee et al., 2014), and spatially clustered coefficient models (Li & Sang, 2019). Spatial filtering (SF) methods involve regressing a covariate (attribute variable) on synthetic variates, which account for spatial autocorrelation in the surface of the variables by using Moran's *I* or a similar measure of spatial autocorrelation. Eigenvector spatial filtering (ESF) follows the observation that eigenvectors of appropriately weighted spatial connectivity matrices based on Moran's *I* (or an equivalent measure), represent a "kaleidoscope . . . of map patterns" (Griffith, 2003, p. 142), which are exhaustive of all the possible patterns within the geographic arrangement of observations. The spatial pattern of any variable can hence be approximated by some linear combination of this set of patterns and be separated from the aspatial part. Griffith (2008) extended this basic linear model by interacting predictors with selected eigenvector terms to enable estimation of spatially varying parameter estimates, hence *localizing* the modeling approach. A subsequent development alternatively uses the map patterns from ESF to localize a model by parameterizing the covariance of random effects that can similarly be combined with predictors to create local estimates (Murakami et al., 2017). Methods within the Bayesian SVC framework use a Bayesian hierarchical model specification, which closely resembles a correlated mixed-effects model. The global mean response effect in the model is estimated at the lower level of the hierarchy whereas the site-specific parameter estimates are modeled at the uppermost level (Gelfand et al., 2003). Of these techniques, SVC and local ESF are "full-map" techniques in that they use all available data in a single regression specification to obtain a set of site-specific parameter estimates (Gelfand et al., 2003; Griffith, 2008). SVC models are commonly estimated using Markov Chain Monte Carlo algorithms, and their implementation is generally more complex than GWR models.

Although these local modeling frameworks differ in how they represent spatially varying processes, they share the basic idea that processes *might* vary over space and that global model representations such as those in equations (1.1) and (1.2) are inadequate in such circumstances. Also common to all four frameworks is the assumption of *process spatial dependency*. That is, if processes do vary over space, they are unlikely to vary randomly but instead are likely to follow a process-based equivalent to Tobler's Law such that "*the processes leading to a particular outcome will be more similar in nearby places than in more distant ones*". Comparisons of different forms of local models can be found elsewhere (Harris, 2019; Murakami et al., 2019; Wolf et al., 2018; and Oshan & Fotheringham, 2018). However, preliminary results suggest, reassuringly, that the frameworks provide substantially similar results in terms of local parameter estimates, although this is a thriving research area and further comparative work is warranted. In the remainder of this book, we concentrate on the MGWR framework largely because of its familiarity and extensive use in the literature, but also because it is arguably clearer conceptually how the model captures locally varying behavior and because the framework is more extendable and scalable. As demonstrated in Chapter 5, freely available, user-friendly software exists for MGWR, which makes model calibration accessible and reproducible.

We now examine why some processes, especially those related to the behavior of human beings, might vary over space and, consequently, why local models should be preferred to global ones in the modeling of such behavior.

1.3 Why "Local"—the Role of Spatial Context

The raison d'être of local models is that the processes being modeled *might* vary across space. Consequently, it is pertinent to ask how and why processes could vary over space and which processes might be susceptible to such variation. Clearly, there is no reason for many processes to be anything other than constant across space—for example, E does not equal $mc^{1.9}$ in some places and $mc^{2.1}$ in others. So which processes might exhibit spatial variability and why? The answers to these questions are unfortunately not definitive, and it is impossible to state *a priori* that any given processes will exhibit spatial heterogeneity. However, there is a substantial amount of empirical evidence that supports the notion that many processes related to human behavior do vary over space, and there is a vast literature, both theoretical and empirical, which suggests that 'place matters' and that local 'context' can have a major impact on people's beliefs, preferences, and actions (*inter alia*, Agnew, 2014; Duncan & Savage, 2006; Golledge, 1997; Goodchild, 2011; Gould, 1991; Hartshorne, 1939a, 1939b; Harvey & Wardenga, 2006; Pred, 1984; Relph, 1976; Sayer, 1985; Thomae, 1999; Tuan, 1979; Winter et al., 2009; Winter & Freksa, 2012). As Enos (2017, p. 78) states:

> *Context—or, more precisely, social geography—can directly affect our behavior and is therefore tremendously important.*

So how might location affect behavior? A link between place and behavior can arise, among other reasons, if people are influenced by the people they talk to on a regular basis, or by their local media, or if they face long-term conditions that are peculiar to certain locales and which shape people's outlooks on issues. Evidence for this on a large scale can be seen in geographic variations in preferences for certain types of foods, music, house styles, political parties, and so on (Walker & Li, 2007; Enos, 2017; Escobar, 2001; Anderson & West, 2006; Hudson, 2006; Shortridge, 2003; Agnew, 1996; Braha & Aguiar, 2017; Fotheringham et al., 2021). On a more local scale, there are a number of reasons for suspecting that location may have an influence on behavior. These can be summarized as follows:

1. Traditions, persistent adverse or beneficial conditions, customs, lifestyles, and psychological profiles common to an area can affect social norms, which in turn affect individual behavior. Several studies have commented on personality differences across regions and how these can explain behavioral differences. Krug and Kulhavy (1973 p. 73), for example, state regarding the United States:

   > *It is clear that practically significant personality differences do exist across the country in a measurable and quantifiable way.*

   Similarly, Rentfrow et al. (2015, p. 1) state:

   > *Recent investigations indicate that personality traits are unevenly distributed geographically . . . (these) are associated with a range of important political, economic, social and health outcomes.*

   In a separate study, Rentfrow et al. (2013 p. 996) report:

   > *Characterizations of regions based on the psychological characteristics of the people who live in them are appealing because psychological factors are likely to be the driving forces behind the individual-level behaviors that eventually get expressed in terms of macrolevel social and economic indicators.*

The argument in each of the above studies is that there is something inherent in the psychological profiles of residents of different locations, which leads them to react differently to similar stimuli. For instance, many people in the upper Midwest of the United States can trace their ancestry back to Scandinavia where a "*private deprivation for the public good*" spirit is fairly common, whereas in other parts of the country a feeling of self-reliance and self-governance is more common. These traits, which transcend individual demographic characteristics, can manifest themselves in a variety of ways, such as how people feel about taxation, how they vote, and the lifestyles they lead.

2. Local media and selective news representation. Several commentators have noted the influence of the news media on the behavior of individuals (*inter alia*, DellaVigna & Kaplan, 2007; Beck et al., 2002; Garz, 2018; Hollanders & Vliegenthart, 2011). Increasingly few people read neutral media and the slanted view they receive can have a strong influence on both what they believe and how they behave, leading to spatial variations in behavior that are independent of personal characteristics. This phenomenon is growing and Bishop (2009, p. 74) claims that we live in "*gated media communities*" insofar as we only engage with media that support our views. This leads to a situation where objectivity is diminished and people rarely change their views. Indeed, initial views often become hardened over time: even when people hear debates they tend to only listen to the arguments that support their existing views, especially when they are in the company of like-minded individuals, a trait known as "*confirmation bias*". The massive expansion of information outlets through social media and the internet in general has only served to further separate people and harden views, which can become extreme (such as those who follow QAnon sites, for example).
3. The influence of friends, family, and local organizations. This is perhaps the most obvious cause of spatial contextual influences on behavior and is referred to as "*social imitation*"—the desire to fit in with people around us. That is, who we talk to regularly, at home, at work, at social gatherings, or in the street, can sway our opinions and values leading to shared behavioral traits linked to location (Beck et al., 2002; Huckfeldt & Sprague, 1995; Huckfeldt et al., 1995). This is amplified by what social psychologists refer to as "*group polarization*"—over time, groups become more extreme in the direction of the average opinion of individual group members. This can occur for several reasons such as individuals not wanting to stand out from the group; hearing ideas frequently increases the belief that they are correct and are less likely to be questioned; being more extreme in one's opinions brings approbation from the group; and individuals with minority opinions become less likely to air such views so that debate and contradictory opinions become rare.
4. Environmental conditions. Several authors have suggested that the environment can affect behavior such as people in warmer parts of the country

behaving differently from people in colder regions (Anderson, 1987; Gastil, 1975; Zelinsky, 1973). However, it is more difficult to see how environmental determinants could account for the micro-contextual effects on behavior we often measure.

Whatever combination of the above factors is responsible for people's values and actions being influenced by where they live, this is amplified by selective migration and the tendency of people to seek out like-minded individuals (homophily) or avoid people with dissimilar views (xenophobia), concepts that have been well documented and researched (Bishop, 2009; Borchert, 1972; Zelinsky, 1973; Schelling, 1971; Sakoda, 1971). This is seen very clearly by the paradox in US Presidential elections where the overall vote is often evenly split between Republicans and Democrats, but the majority of people live in neighborhoods where the split in the vote is very uneven.

Despite a wealth of evidence that place matters and that location can help shape preferences and actions, it could be argued that what is referred to as 'context' is merely a catch-all term for those covariates not included in the model either because they have not been conceived of having importance or because they are difficult to measure (Hauser, 1970; King, 1996; McAllister, 1987). Even though many sociological and psychological studies have pointed to the relevance of context (Beck et al., 2002; Enos, 2017; Oreg & Katz-Gerro, 2006; Plaut et al., 2002; Rentfrow et al., 2015; Krug & Kulhavy, 1973) and a great number of studies have espoused the role of location in affecting behavior from a theoretical viewpoint (Blake, 2001; Books & Prysby, 1998; Chandola et al., 2005; Rousseau & Fried, 2001; Snedker et al., 2009), it could be claimed that whatever the effects of location are, they could, theoretically, be measured and incorporated into the model. However, there are two counterarguments to such a claim.

The first is that this claim relates to a theoretical construct, and in practice, we never have the luxury of both knowing and being able to measure all the relevant variables that affect a person's behavior. Whether context is a 'real' effect or simply a catch-all for variables that cannot be or have not been measured will remain elusive and is arguably somewhat irrelevant. Whatever its source, the ability to capture a 'context' effect within local models is better than not accounting for it at all, as is the case in traditional global models. In global models, the omission of an important explanatory variable will create misspecification bias in the parameter estimates associated with any covariate that has some degree of covariance with the omitted variable (for an example of this, and the calculation of the explicit degree of misspecification bias caused by an omitted variable, see Fotheringham [1983a, 1984]). Equally, if the processes being modeled do vary over space but they are modeled with a traditional global model, the latter will be grossly misspecified, and the estimates of the parameters derived from a global model will simply represent the average of many different processes. As such, the global parameter estimates will be as unrepresentative of the spatial varying processes as is, for example, the mean annual rainfall of the United States as a measure of the annual rainfall in each county.

The second argument (see Figure 1.2) is that spatial context can influence behavior in *two* ways and that much of the debate regarding the role of context has arisen

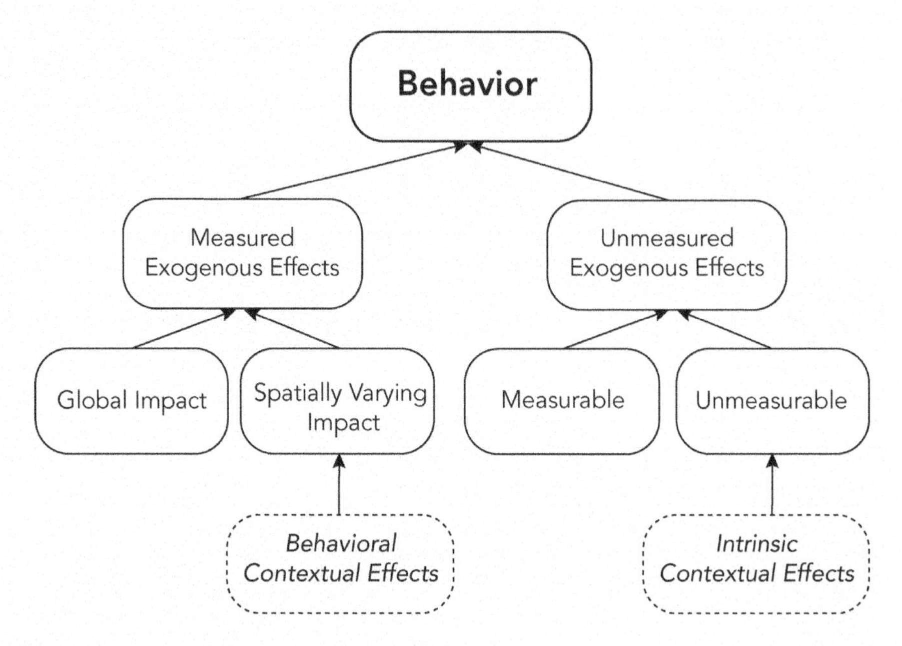

FIGURE 1.2
On the Roles of Context in Determining Behavior.

because there has been either confusion over these two roles or ignorance of one of them. Suppose that we construct a model that relates some aspect of human behavior to a set of attributes we think may influence this behavior. These influences can be divided into those effects we have measured and included in our model and those we have not. Unmeasured effects are those we have not included in our model for one of two reasons: we have not thought to include them (the 'measurable' unmeasured effects); or we cannot measure them (the 'unmeasurable' unmeasured effects). Ideally, we want to minimize (set to zero) the measurable unmeasured effects, and we strive to do this by giving a great deal of thought to model construction and variable selection. However, we recognize that in many situations there are some effects, which we cannot possibly measure. In models of spatial processes, these represent the intangible influences of location and what we call here "*Intrinsic Contextual Effects*". These are the contextual effects that King (1996) and others claim that as scientists we should strive to eliminate, a goal that is both admirable and often unattainable. What we can do is to try to remove all the unmeasured, measurable effects in our models, but, inevitably, some effects will remain as unmeasurable.

There is, however, also a second type of contextual effect, termed here "*behavioral contextual effects*", which relate to the influence of location on how the measured effects in the model affect behavior. So, even if we were to include in our model all possible influences on behavior, and hence eliminate intrinsic contextual effects, behavioral contextual effects could still be important if any of the measured exogenous effects has a spatially varying impact on behavior. For instance, being unemployed in one part of a city may be a greater indicator of committing a crime than in another part. Equally, behavioral contextual effects would occur if young voters, *ceteris paribus*, had a preference for political party A in one region but a preference for political party B in a different region. Beck et al. (2002, p. 69) comment on this view of context in US voting behavior:

> *American voters do not operate in the social vacuum that much of the contemporary voting literature seems to assume. Rather, voters' enduring personal characteristics interact with the messages they are receiving from the established social context in which they operate. This context cannot be ignored in trying to understand voting and electoral outcomes in any election.*

The distinction we make here in the two ways context can influence behavior is important for what follows in the remainder of this book because intrinsic contextual effects are captured in local models by the local intercept whereas behavioral contextual effects are captured by locally varying slope parameters.

Despite the common acceptance that context can and often does affect people's behavior and that the effects of context will vary by location, there remain several questions about its role in determining behavior. As Enos (2017, p. 120) states:

> *Nobody doubts that context can affect behavior and careful studies of "neighborhood effects" have strongly suggested it can. However, the exact nature of contextual effects—how much they really matter—is elusive to researchers.*

This sentiment is echoed by O'Loughlin (2018, p. 148):

> *But if context has remained a mantra in political geography, how do we measure its importance?*

and by Braha and de Aguiar (2017, p. 1):

> *The question of how to separate and measure the effect of social influence is therefore a major challenge for understanding collective human behavior.*

As we shall see in the remainder of this book, local models provide a means both of separating the effects of intrinsic contextual effects from the confounding effects of socioeconomic factors and also of measuring the importance of contextual effects in determining behavior, both of which have eluded social scientists until recently. In summary, although one cannot predict which processes might be subject to spatial heterogeneity, certainly in modeling any aspect of human behavior, the strategic view would be to assemble as comprehensive a set of covariates as possible and to calibrate a local model using these covariates. The local model will be able to capture both intrinsic and behavioral contextual effects if they are present and will tend to indicate that a global model is appropriate in the situation where the processes are stationary over space.

1.4 A Conceptual Overview of MGWR

In the calibration of a global model, such as those in equations (1.1) and (1.2), with spatial data recorded at a number of locations, the typical procedure would involve using the data on $y, x_1, x_2, \ldots x_k$ recorded at each location in a single calibration, which would yield one estimate of each parameter in the model. A local model, such as that depicted in equation (1.3), allows the estimation of location-specific parameters, and in the MGWR framework a separate model is calibrated for each location. If each location had a sufficiently large number of observations on $y, x_1, x_2, \ldots x_k$, then it would be straightforward to undertake such calibrations as the global model could be calibrated separately at every location. Each location-specific calibration would thus be based only on data obtained at that specific location, and the parameter estimates would therefore reflect only the processes taking place at that location. Indeed, if we had the luxury of large numbers of observations at each location, why would we merge data from other locations? We only do this because typically we have an insufficient number of observations (usually only one) on $y, x_1, x_2, \ldots x_k$ at each location to calibrate a location-specific model. So how do we calibrate local models such as that in equation (1.3) when we only have one set of measurements of $y, x_1, x_2, \ldots x_k$ at each location?

The answer is, we borrow data from nearby locations. However, this immediately prompts the further question: "*Which nearby locations?*". To understand the logic of how this question is answered in MGWR, consider the two surfaces of parameter values depicted in Figures 1.3 (a) and 1.3 (b), both of which have sets of locations marked at which data on $y, x_1, x_2, \ldots x_k$ are available. A local model calibration is to be undertaken at location i in both figures. In Figure 1.3 (a) there is no variation in the parameter surface so that data can be borrowed from every location without introducing any bias into the local parameter estimate for i. Indeed, the 'local' calibration of the model would be identical for every location since each calibration would use all the available data and would be equivalent to a global model. In Figure 1.3 (b), however, the parameter surface exhibits spatially dependent variation

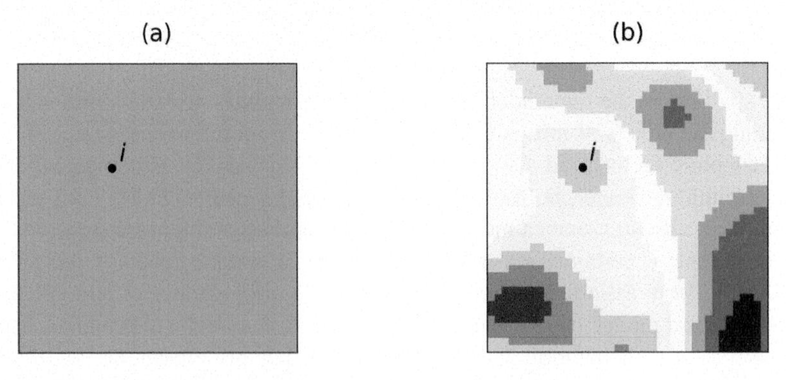

FIGURE 1.3
Two Sets of Local Parameters: (a) Constant Over Space and (b) Spatially Varying.

so that locations in closer proximity to each other generally have more similar parameter values than do locations that are further apart. In this case, bias will be introduced into the local calibration at i by borrowing data from other locations because the process being modeled varies over space. This bias will tend to increase the further away are the locations from which data are borrowed. Consequently, it might be preferable in this situation to give less weight to data from locations, which are far from location i in the local calibration. The weighting can be controlled by a distance-decay parameter, or bandwidth, so that the weights range between 1 (at the local calibration point) and 0 (at the bandwidth). This process is summarized in Figure 1.4 and allows parameter estimates to become spatially varying because the data used in each local calibration are specific to that location.

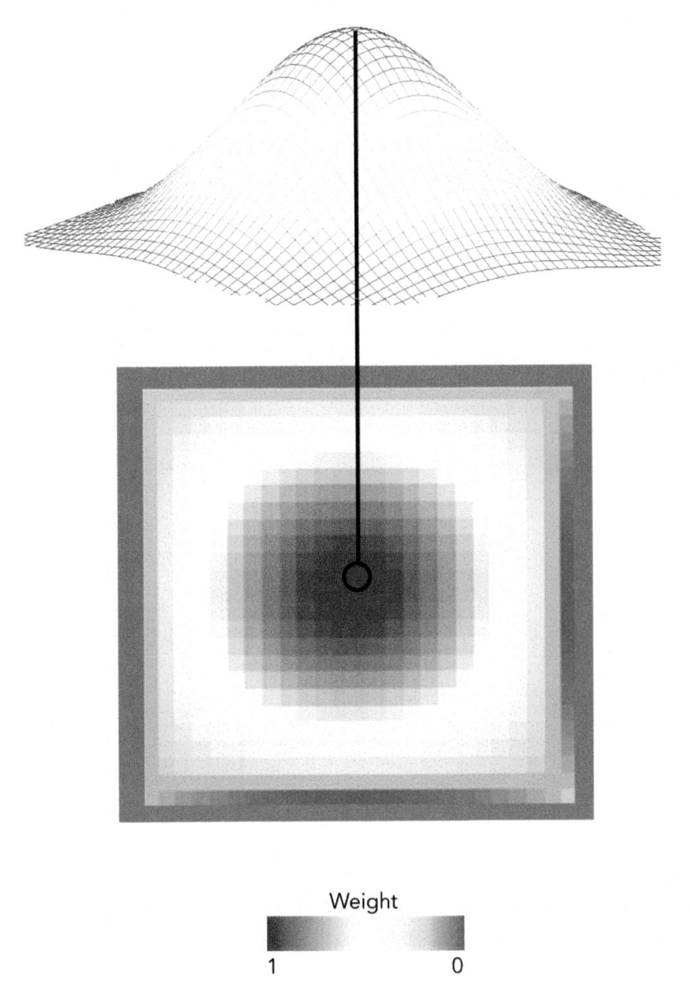

FIGURE 1.4
Spatial Weighting in Geographically Weighted Regression.

Determining the optimal bandwidth is thus an important element of an M(GWR) calibration and involves a trade-off between the amount of bias in the local parameter estimates and their uncertainty (or variance). As discussed earlier, bias in the local parameter estimates is caused by borrowing data from other locations where the processes that produced those data may not be the same as those that are being estimated at the regression location. Uncertainty in the local parameter estimates exists because we are using a sample of data from which to calculate these estimates, and this uncertainty will decrease as more data are borrowed in each local regression. Hence, small bandwidths generate local parameter estimates with lower bias but increased uncertainty whereas large bandwidths generate local parameter estimates with greater bias but lower uncertainty. This situation is captured in Figure 1.5. Note that the range of data borrowing for the local regression at i, the bandwidth, can be estimated in terms of a physical distance or in terms of the number of nearest neighbors. It is generally more intuitive, and eases comparisons across studies, if the latter is used.

Initially, as data are borrowed from more distant locations and added to the local regressions, the reduction in parameter uncertainty outweighs the increase in parameter bias until a point is reached (the optimal bandwidth) at which the increase in

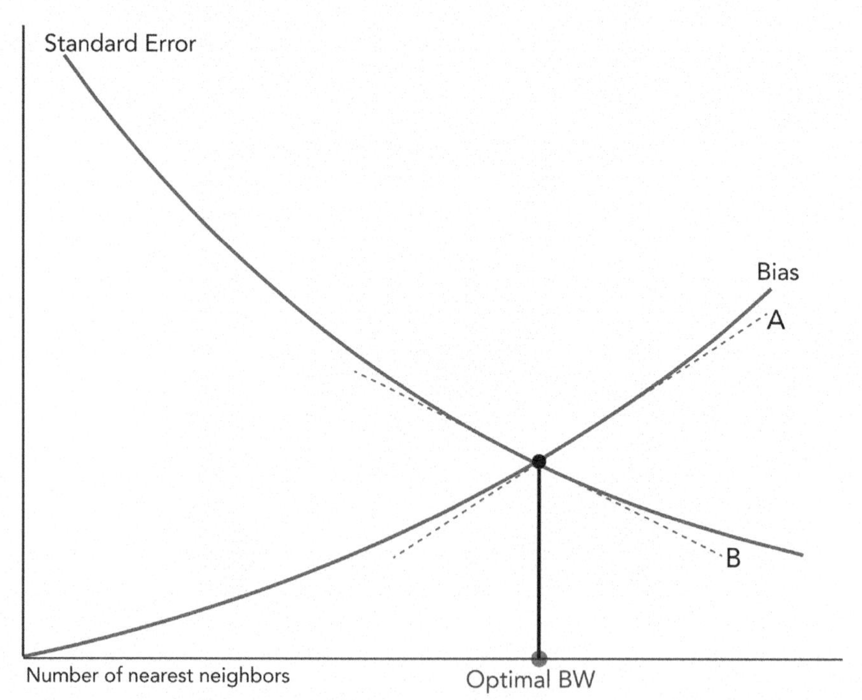

FIGURE 1.5
Optimal Bandwidth Selection in GWR.

parameter bias outweighs the reduction in uncertainty. If the process being modeled has a high degree of spatial variability, the bandwidth will be small; if the process has low spatial variability, the bandwidth will be large. A global process will result in an infinitely large bandwidth so that a global model is therefore a special case of a local model in which all the processes being modeled have infinitely large bandwidths, giving a weight of 1 to all data points. Consequently, the optimized bandwidth provides useful information on the spatial scale over which a process varies, a point to which we return in Chapter 4. Given that different processes may exhibit different degrees of variation over space, we need some comparative measure of their relative variability, and for this we chose the point at which the amount of bias in the local parameter estimates increases at a faster rate than the decrease in the parameter uncertainty (for more details, refer to Yu et al., 2020a). Hence, the optimal bandwidth denotes the limit where the data borrowed for the local regression reduce parameter uncertainty more than they increase parameter bias. Beyond the bandwidth, the addition of further data to the regression would increase bias more than it would decrease uncertainty. This is a measure of a property of spatially varying processes, which can be compared both between different processes and across applications to inform on the relative spatially varying nature of local processes.

Another way of viewing the conceptual basis for local modeling with MGWR is to think of data being composed of '*information*' and '*misinformation*' in terms of the processes operating at location *i*. Data close to location *i*, and which presumably are the product of processes similar to those being estimated for *i*, contain more information than misinformation and so are included in the local regression. Data far from *i*, and which are likely to be the product of different processes to those at *i*, contain more misinformation than information on the processes at *i* and so are excluded from the location regression at *i*. The bandwidth is thus the point in space from *i* beyond which data contain more misinformation than information about the processes at *i*. The bandwidth will clearly be a function of how much spatial variability is exhibited by a process—the more spatial variation in a process, the smaller will be the bandwidth. This interpretation of the role of the bandwidth is not dissimilar to that of the range in kriging but in the case of local modeling the focus is on *process* variability rather than *data* variability.

The bandwidth within MGWR has an interpretable real-world meaning and can be either a distance-based measure or the number of nearest neighbors used in each local regression (see Chapter 2). For example, the interpretation of a bandwidth of 100 nearest neighbors is that the process being estimated at location *i* is affected by other neighboring locations in a spatially discounted manner up to the 100th nearest neighbor of *i*. Data at all other locations have no influence on the local regression at *i*. If the bandwidth is measured as a distance, the reported bandwidth in MGWR calibrations is the distance at which the weight of a data point falls below 0.05 and is termed the *effective bandwidth*. More detail on bandwidths is provided in Chapters 3 and 4.

The predecessor of MGWR, GWR, has the limitation that the same bandwidth is assumed to apply to each relationship in the model. This is clearly an unreasonable assumption in most situations. As Harvey (1967, pp. 71–72) noted "*different processes*

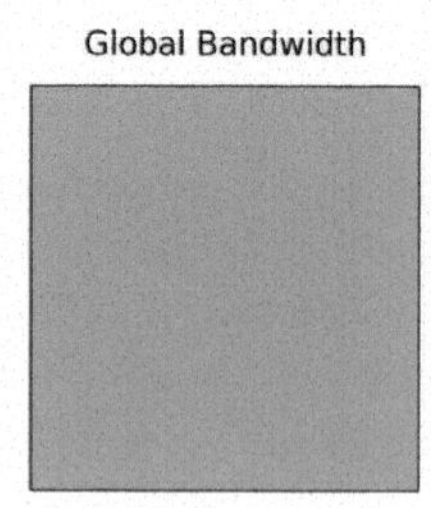

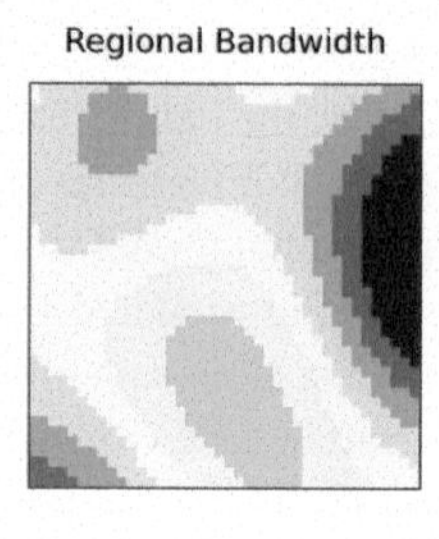

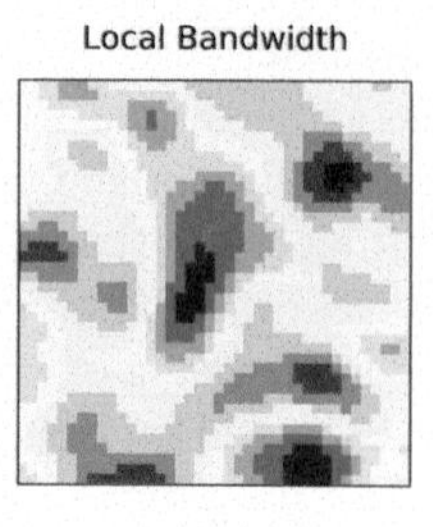

FIGURE 1.6
Four Local Parameter Surfaces and Their Associated Optimal Bandwidths.

become significant to our understanding of spatial patterns at different scales", an observation expounded upon decades later by Manley et al. (2006, p. 143):

> *Spatial distributions are based on processes taking place in geographical space. A mapped pattern may reflect several distinct processes, each of which may affect a different area and operate at a different scale. The challenge for the spatial analyst is to identify these processes and evaluate their importance from the spatial pattern observed.*

MGWR overcomes this limitation by reformulating GWR as a generalized additive model (GAM):

$$y_i = \sum_{j=1}^{k} f_{ji} + \varepsilon_i \tag{1.4}$$

where f_{ji} is a smoothing function (i.e., spatial weight or data-borrowing scheme) applied to the jth explanatory variable at location i (Fotheringham et al., 2017; Yu et al., 2020b). This model can then be calibrated using a backfitting algorithm (see Chapter 2 for further details) that derives a separate bandwidth parameter for each of the k processes being modeled as shown in Figure 1.6.

The ability to estimate covariate-specific bandwidths in MGWR not only allows the modeling of spatial heterogeneity across different spatial processes but also provides more accurate inference, minimizes overfitting, moderates concurvity (i.e., collinearity due to similar functional transformations), and reduces bias in the coefficient estimates (Fotheringham et al., 2017; Wolf et al., 2018; Oshan et al., 2019a, 2019b; Li et al., 2020; Yu et al., 2020a, 2020b). All of these issues are discussed in subsequent chapters.

1.5 Summary

The premise of this book is that there are reasons to suspect that the processes producing the data we observe about human beings and their actions may not be stationary due to the influence of spatial context. This is a relatively new perspective in quantitative spatial sciences. Should social processes exhibit spatial heterogeneity, what are the ramifications of this? Should we ignore this possibility and continue calibrating global models, which assume spatial stationarity and yield only average statements of spatially varying processes? This book takes the counterview that we should try to model and explore any spatial process heterogeneity as an interesting and potentially useful facet of human behavior. It further takes the view that if processes do vary over space because of spatial contextual effects, they are unlikely to vary randomly and almost certainly will exhibit some degree of spatial dependency—processes at locations in close proximity are more likely to be similar than are processes at locations farther apart. We can model the degree to which different processes vary over space through MGWR. This local modeling technique produces covariate-specific bandwidths, yielding a comparative measure of the spatial extent that a process is relatively stable. Larger bandwidths indicate processes that are more stable over space, and a global model is an extreme case of a local model with infinitely large bandwidths. We now describe the statistical operation of MGWR in more detail.

Note

1. It should be noted that the physical sciences are not immune from the issue of equifinality as the recent debates on climate change demonstrate.

References

Agnew, J. A. (1996). Mapping politics: How context counts in electoral geography. *Political Geography*, *15*(2), 129–146.

Agnew, J. A. (2014). *Place and politics: The geographical mediation of state and society*. Milton Park: Routledge.

Anderson, C. A. (1987). Temperature and aggression: Effects of quarterly, yearly, and city rates of violent and nonviolent crime. *Journal of Personality and Social Psychology*, *52*, 1161–1173.

Anderson, S. T., & West, S. E. (2006). Open space, residential property values, and spatial context. *Regional Science and Urban Economics*, *36*(6), 773–789.

Anselin, L. (1988). *Spatial econometrics: Methods and models*. Dordrecht: Kluwer.

Anselin, L. (1995). Local indicators of spatial association—LISA. *Geographical Analysis*, *27*(2), 93–115.

Anselin, L. (2002). Under the hood: Issues in the specification and interpretation of spatial regression models. *Agricultural Economics*, *27*(3), 247–267.

Anselin, L. (2009). Spatial regression. In A. S. Fotheringham & P. A. Rogerson (Eds.), *The SAGE handbook of spatial analysis* (pp. 254–275). London: SAGE.

Banerjee, S. (2003). *Hierarchical modeling and analysis for spatial data*. Boca Raton, FL: Chapman & Hall.

Banerjee, S., Carlin, B. P., & Gelfand, A. E. (2014). *Hierarchical modeling and analysis for spatial data*. Boca Raton, FL: Chapman & Hall.

Beck, P. A., Dalton, R. J., Greene, S., & Huckfeldt, R. (2002). The social calculus of voting: Interpersonal, media, and organizational influences on Presidential choices. *American Political Science Review, 96*(1), 57–73.

Bishop, B. (2009). *The big sort—why the clustering of like-minded America is tearing us apart*. Boston, MA: Houghton Mifflin.

Blake, D. E. (2001). Contextual effects on environmental attitudes and behavior. *Environment and Behavior, 33*(5), 708–725.

Books, J., & Prysby, C. (1998). Studying contextual effects on political behavior: A research inventory and agenda. *American Politics Quarterly, 16*(2), 211–238.

Borchert, J. R. (1972). America's changing metropolitan regions. *Annals of the Association of American Geographers, 62*, 352–373.

Braha, D., & de Aguiar, M. A. M. (2017). Voting contagion: Modeling and analysis of a century of U.S. Presidential elections. *PLoS One, 12*(5), e0177970.

Brunsdon, C., Fotheringham, A. S., & Charlton, M. E. (1996). Geographically weighted regression: A method for exploring spatial nonstationarity. *Geographical Analysis, 28*, 281–298.

Chandola, T., Clarke, P., Wiggins, R. D., & Bartley, M. (2005). Who you live with and where you live: Setting the context for health using multiple membership multilevel models. *Journal of Epidemiology and Community Health, 59*(2), 170–175.

Chetty, R., & Hendren, N. (2018). The impacts of neighborhoods on intergenerational mobility I: Childhood exposure effects. *The Quarterly Journal of Economics, 133*(3), 1107–1162.

Cliff, A. D., & Ord, J. K. (1981). *Spatial processes models and applications*. London: Pion.

Darmofal, D. (2008). The political geography of the new deal realignment. *American Politics Research, 36*(6), 934–961.

DellaVigna, S., & Kaplan, E. (2007). The Fox news effect: Media bias and voting. *The Quarterly Journal of Economics, 122*(3), 1187–1234.

Diez-Roux, A. V. (1998). Bringing context back into epidemiology: Variables and fallacies in multilevel analysis. *American Journal of Public Health, 88*(2), 216–222.

Diez-Roux, A. V. (2001). Investigating neighborhood and area effects on health. *American Journal of Public Health, 91*(11), 1783–1789.

Duncan, C., Jones, K., & Moon, G. (1998). Context, composition and heterogeneity: Using multilevel models in health research. *Social Science & Medicine, 46*(1), 97–117.

Duncan, S., & Savage, M. (2006). Space, scale and locality. *Antipode, 21*, 179–206.

Enos, R. D. (2017). *The space between us: Social geography and politics*. Cambridge: Cambridge University Press.

Escobar, A. (2001). Culture sits in places: Reflections on globalism and subaltern strategies of localization. *Political Geography, 20*(2), 139–174.

Fotheringham, A., & O'Kelly, M. E. (1989). *Spatial interaction models: Formulations and applications*. Berlin: Springer.

Fotheringham, A. S. (1983a). Some theoretical aspects of destination choice and their relevance to production-constrained gravity models. *Environment and Planning A, 15*(8), 1121–1132.

Fotheringham, A. S. (1983b). A new set of spatial interaction models: The theory of competing destinations. *Environment and Planning A, 15*(1), 15–36.

Fotheringham, A. S. (1984). Spatial flows and spatial patterns. *Environment and Planning A*, *16*(4), 529–543.
Fotheringham, A. S., Brunsdon, C., & Charlton, M. (2002). *Geographically weighted regression: The analysis of spatially varying relationships*. London: Wiley.
Fotheringham, A. S., Charlton, M. E., & Brunsdon, C. (1996). The geography of parameter space: An investigation of spatial non-stationarity. *International Journal of Geographic Information Systems*, *10*, 605–627.
Fotheringham, A. S., Li, Z., & Wolf, L. J. (2021). Scale, context and heterogeneity: A spatial analytical perspective on the 2016 US Presidential election. *Annals of the American Association of Geographers*, *111*(6), 1602–1621.
Fotheringham, A. S., Yang, W., & Kang, W. (2017). Multiscale geographically weighted regression (MGWR). *Annals of the American Association of Geographers*, *107*(6), 1247–1265.
Garz, M. (2018). Effects of unemployment news on economic perceptions—evidence from German Federal States. *Regional Science and Urban Economics*, *68*, 179–190.
Gastil, R. D. (1975). *Cultural regions of the United States*. Seattle, WA: University of Washington Press.
Gelfand, A. E., Kim, H.-J., Sirmans, C. F., & Banerjee, S. (2003). Spatial modeling with spatially varying coefficient processes. *Journal of the American Statistical Association*, *98*(462), 387–396.
Getis, A. (1991). Spatial interaction and spatial autocorrelation: A cross-product approach. *Environment and Planning A: Economy and Space*, *23*(9), 1269–1277.
Getis, A., & Ord, J. K. (1992). The analysis of spatial association by use of distance statistics. *Geographical Analysis*, *24*(3), 189–206.
Gibbons, S., & Overman, H. G. (2012). Mostly pointless spatial econometrics? *Journal of Regional Science*, *52*(2), 172–191.
Golledge, R. G. (1997). *Spatial behavior: A geographic perspective*. New York: Guilford Press.
Goodchild, M. F. (2011). Formalizing place in geographic information systems. In L. M. Burton, S. A. Matthews, M. Leung, S. P. Kemp, & D. T. Takeuchi (Eds.), *Communities, neighborhoods, and health: Expanding the boundaries of place* (pp. 21–33). New York: Springer.
Gould, P. (1991). On reflections on Richard Hartshorne's the nature of geography. *Annals of the Association of American Geographers*, *81*(2), 328–334.
Griffith, D. A. (2003). *Spatial autocorrelation and spatial filtering: Gaining understanding through theory and scientific visualization*. New York: Springer Science & Business Media.
Griffith, D. A. (2008). Spatial-filtering-based contributions to a critique of geographically weighted regression (GWR). *Environment and Planning A*, *40*(11), 2751–2769.
Griffith, D. A., & Csillag, F. (1993). Exploring relationships between semi-variogram and spatial autoregressive models. *Papers in Regional Science*, *72*(3), 283–295.
Haining, R. P. (1993). *Spatial data analysis in the social and environmental sciences*. Cambridge: Cambridge University Press.
Haining, R. P. (2003). *Spatial data analysis: Theory and practice*. Cambridge: Cambridge University Press.
Harris, P. (2019). A simulation study on specifying a regression model for spatial data: Choosing between heterogeneity and autocorrelation effects. *Geographical Analysis*, *51*, 151–181.

Hartshorne, R. (1939a). The nature of geography: A critical survey of current thought in the light of the past. *Annals of the Association of American Geographers*, *29*(3), 173–412.
Hartshorne, R. (1939b). The character of regional geography. *Annals of the Association of American Geographers*, *29*(4), 436–456.
Harvey, D. W. (1967). Pattern, process, and the scale problem in geographical research. *Transactions of the Institute of British Geographers*, *45*, 71–78.
Harvey, F., & Wardenga, U. (2006). Richard Hartshorne's adaptation of Alfred Hettner's system of geography. *Journal of Historical Geography*, *32*(2), 422–440.
Hauser, R. M. (1970). Context and consex: A cautionary tale. *American Journal of Sociology*, *75*(4, Part 2), 645–664.
Hay, A. M., & Johnston, R. J. (1983). The study of process in quantitative human geography. *Espace Geographique*, *12*(1), 69–76.
Hollanders, D., & Vliegenthart, R. (2011). The influence of negative newspaper coverage on consumer confidence: The Dutch case. *Journal of Economic Psychology*, *32*(3), 367–373.
Hubert, L. J., Golledge, R. G., & Costanzo, C. M. (1981). Generalized procedures for evaluating spatial autocorrelation. *Geographical Analysis*, *13*(3), 224–233.
Huckfeldt, R., Beck, P. A., Dalton, R. J., & Levine, J. (1995). Political contexts, cohesive social groups, and the communication of public opinion. *American Journal of Political Science*, *39*, 1025–1054.
Huckfeldt, R., & Sprague, J. (1995). *Citizens, politics and social communication*. New York: Cambridge University Press.
Hudson, R. (2006). Regions and place: Music, identity and place. *Progress in Human Geography*, *30*(5), 626–634.
Kelejian, H. H., & Prucha, I. R. (1998). A generalized spatial two-stage least squares procedure for estimating a spatial autoregressive model with autoregressive disturbances. *The Journal of Real Estate Finance and Economics*, *17*(1), 99–121.
Kelejian, H. H., & Prucha, I. R. (1999). A generalized moments estimator for the autoregressive parameter in a spatial model. *International Economic Review*, *40*(2), 509–533.
King, G. (1996). Why context should not count. *Political Geography*, *15*, 159–164.
Krug, S. E., & Kulhavy, R. W. (1973). Personality differences across regions of the United States. *The Journal of Social Psychology*, *91*(1), 73–79.
Lesage, J. P. (2016). Bayesian estimation of spatial autoregressive models. *International Regional Science Review*, *20*(1–2), 113–129.
Lesage, J. P., & Pace, R. K. (2010). Spatial econometric models. In M. M. Fischer & A. Getis (Eds.), *Handbook of applied spatial analysis: Software tools, methods and applications* (pp. 355–376). Berlin: Springer.
Li, F., & Sang, H (2019). Spatial homogeneity pursuit of regression coefficients for large datasets. *Journal of the American Statistical Association*, *114*, 1050–1062.
Li, Z., Fotheringham, A. S., Oshan, T. M., & Wolf, L. J. (2020). Measuring bandwidth uncertainty in multiscale geographically weighted regression using Akaike weights. *Annals of the American Association of Geographers*, *110*(5), 1500–1520.
Lloyd, C. D. (2007). *Local models for spatial analysis*. Boca Raton, FL: CRC Press.
Manley, D., Flowerdew, R., & Steel, D. G. (2006). Scales, levels and processes: Studying spatial patterns of British census variables. *Computers, Environment and Urban Systems*, *30*(2), 143–160.
Matheron, G. (1963). Principles of geostatistics. *Economic Geology*, *58*(8), 1246–1266.
McAllister, I. (1987). II. Social context, turnout, and the vote: Australian and British comparisons. *Political Geography Quarterly*, *6*(1), 17–30.

Moran, P. A. P. (1948). The interpretation of statistical maps. *Journal of the Royal Statistical Society. Series B (Methodological)*, *10*(2), 243–251.
Moran, P. A. P. (1950). Notes on continuous stochastic phenomena. *Biometrika*, *37*(1/2), 17–23.
Murakami, D., Lu, B., Harris, P., Brunsdon, C., Charlton, M. E., Nakaya, T., & Griffith, D. A. (2019). The importance of scale in spatially varying coefficient modeling. *Annals of the American Association of Geographers*, *109*(1), 50–70.
Murakami, D., Yoshida, T., Seya, H., Griffith, D. A., & Yamagata, Y. (2017). A Moran coefficient-based mixed effects approach to investigate spatially varying relationships. *Spatial Statistics*, *19*, 68–89.
O'Loughlin, J. (2018). Thirty-five years of political geography and political geography: The good, the bad and the ugly. *Political Geography*, *65*, 143–151.
Olsson, G. (1970). Explanation, prediction and meaning variance: An assessment of distance interaction models. *Economic Geography*, *46*, 223–233.
Ord, J. K., & Getis, A. (1995). Local spatial autocorrelation statistics: Distributional issues and an application. *Geographical Analysis*, *27*(4), 286–306.
Ord, J. K., & Getis, A. (2001). Testing for local spatial autocorrelation in the presence of global autocorrelation. *Journal of Regional Science*, *41*(3), 411–432.
Oreg, S., & Katz-Gerro, T. (2006). Predicting pro-environmental behavior cross-nationally: Values: The theory of planned behavior, and value-belief-norm theory. *Environment and Behavior*, *38*(4), 462–483.
Oshan, T. M., Li, Z., Kang, W., Wolf, L. J., & Fotheringham, A. S. (2019a). MGWR: A python implementation of multiscale geographically weighted regression for investigating process spatial heterogeneity and scale. *ISPRS International Journal of Geo-Information*, *8*(6), 269.
Oshan, T. M., Wolf, L. J., Fotheringham, A. S., Kang, W., Li, Z., & Yu, H. (2019b). A comment on geographically weighted regression with parameter-specific distance metrics. *International Journal of Geographical Information Science*, *33*(7), 1289–1299.
Oshan, T. M., & Fotheringham, A. S. (2018). A comparison of spatially varying regression coefficient estimates using geographically weighted and spatial-filter-based techniques. *Geographical Analysis*, *50*(1), 53–75.
Paelinck, J. H. P., & Klaassen, L. H. (1979). *Spatial econometrics*. Aldershot: Gower.
Plaut, V. C., Markus, H. R., & Lachman, M. E. (2002). Place matters: Consensual features and regional variation in American well-being and self. *Journal of Personality and Social Psychology*, *83*(1), 160–184.
Pred, A. (1984). Place as historically contingent process: Structuration and the time-geography of becoming places. *Annals of the Association of American Geographers*, *74*(2), 279–297.
Relph, E. C. (1976). *Place and placelessness*. London: Pion.
Rentfrow, P. J., Gosling, S. D., Jokela, M., Stillwell, D. J., Kosinski, M., & Potter, J. (2013). Divided we stand: Three psychological regions of the United States and their political, economic, social and health correlates. *Journal of Personality and Social Psychology*, *105*(6), 996–1012.
Rentfrow, P. J., Jokela, M., & Lamb, M. E. (2015). Regional personality differences in Great Britain. *PLoS One*, *10*(3), e0122245.
Rousseau, D. M., & Fried, Y. (2001). Editorial: Location, location, location: Contextualizing organizational research. *Journal of Organizational Behavior*, *22*(1), 1–13.
Sakoda, J. (1971). The checkerboard model of social interaction. *Journal of Mathematical Sociology*, *1*, 119–132.

Sampson, R. J. (2019). Neighborhood effects and beyond: Explaining the paradoxes of inequality in the changing American metropolis. *Urban Studies*, *56*(1), 3–32.
Sayer, A. (1985). The difference that space makes. In D. Gregory & J. Urry (Eds.), *Social relations and spatial structures* (pp. 49–66). London: Macmillan Education.
Schelling, T. (1971). Dynamic models of segregation. *Journal of Mathematical Sociology*, *1*, 143–186.
Shortridge, B. G. (2003). A food geography of the Great Plains. *Geographical Review*, *93*(4), 507–529.
Snedker, K. A., Herting, J. R., & Walton, E. (2009). Contextual effects and adolescent substance use: Exploring the role of neighborhoods. *Social Science Quarterly*, *90*(5), 1272–1297.
Thomae, H. (1999). The nomothetic-idiographic issue: Some roots and recent trends. *International Journal of Group Tensions*, *28*(1), 187–215.
Tiefelsdorf, M. (2000). *Modelling spatial processes: The identification and analysis of spatial relationships in regression residuals by means of Moran's I*. Berlin: Springer-Verlag.
Tobler, W. R. (1970). A computer movie simulating urban growth in the Detroit region. *Economic Geography*, *46*, 234.
Tuan, Y.-F. (1979). Space and place: Humanistic perspective. In S. Gale & G. Olsson (Eds.), *Philosophy in geography* (pp. 387–427). Berlin: Springer.
Walker, J. L., & Li, J. (2007). Latent lifestyle preferences and household location decisions. *Journal of Geographical Systems*, *9*(1), 77–101.
Waller, L. A., & Gotway, C. A. (2004). *Applied spatial statistics for public health data*. London: Wiley.
Whittle, P. (1954). On stationary processes in the plane. *Biometrika*, *41*(3/4), 434–449.
Wilson, A. G. (1971). A family of spatial interaction models, and associated developments. *Environment and Planning A*, *3*(1), 1–32.
Winter, S., & Freksa, C. (2012). Approaching the notion of place by contrast. *Journal of Spatial Information Science*, *5*, 31–50.
Winter, S., Kuhn, W., & Krüger, A. (2009). Guest Editorial: Does place have a place in geographic information science? *Spatial Cognition & Computation*, *9*(3), 171–173.
Wolf, L. J., Oshan, T. M., & Fotheringham, A. S. (2018). Single and multiscale models of process spatial heterogeneity. *Geographical Analysis*, *50*(3), 223–246.
Yu, H., Fotheringham, A. S., Li, Z., Oshan, T. M., Kang, W., & Wolf, L. J. (2020b). Inference in multiscale geographically weighted regression. *Geographical Analysis*, *52*(1), 87–106.
Yu, H., Fotheringham, A. S., Li, Z., Oshan, T. M., & Wolf, L. J. (2020a). On the measurement of bias in geographically weighted regression models. *Spatial Statistics*, *38*, 100453.
Zelinsky, W. R. (1973). *The cultural geography of the United States*. Englewood Cliffs, NJ: Prentice Hall.

2

MGWR

The Essentials

2.1 Introduction

In this chapter we outline the basic formulations and operations that constitute MGWR. However, given that MGWR builds on its predecessor, GWR, it is useful to first outline the mechanics of GWR and then describe how and where MGWR differs from GWR. Consequently, the bulk of this chapter is devoted to the fundamental concepts inherent in GWR, and then further developments are described, which make MGWR a significant advance.

2.2 Geographically Weighted Regression

In the previous chapter, conventional "global" models, such as the standard linear regression model, which assume that relationships are constant across a study area, were introduced. As a point of departure, the standard linear regression model is introduced again here in the spatial setting:

$$y_i = \beta_0 + \beta_1 x_{1i} + \beta_2 x_{2i} + \ldots + \beta_k x_{ki} + \varepsilon_i \tag{2.1}$$

where i is a function of coordinates $\{u, v\}$ for a given location in the study area and $i = \{1, 2, 3, \ldots, m\}$, y_i is the observation of the dependent variable at the ith location, β_0 is the intercept, x_{ki} is the observation of the kth explanatory variable at the ith location, β_k is the kth parameter, and ε_i is a random error associated with location i. When using an ordinary least squares (OLS) estimation technique to calibrate the model denoted by equation (2.1), a parameter vector β is estimated by the following calculation in matrix form:

$$\widehat{\beta} = [\boldsymbol{X}^T \boldsymbol{X}]^{-1} \boldsymbol{X}^T \boldsymbol{y} \tag{2.2}$$

DOI: 10.1201/9781003435464-2

where $\boldsymbol{X}$ is an $m \times k$ design matrix of explanatory variables, $\boldsymbol{y}$ is an $m \times 1$ vector of dependent variable observations, and T denotes the transpose operator. Note that the resulting parameter estimates $\widehat{\beta} = \{\widehat{\beta}_0, \widehat{\beta}_1, \widehat{\beta}_2, \ldots, \widehat{\beta}_k\}$ in equation (2.2) are not indexed by i because the sign and magnitude of each represent the relationship between an explanatory variable and the dependent variable *for all of locations in the study area*. The model therefore reinforces the assumption that relationships are constant across space (i.e., stationary) and is unable to accommodate the alternative assumption that relationships may vary across space (i.e., nonstationary). This can lead to model misspecification whenever there is spatial variation in the processes generating the data because the parameter estimates for the global model generally become coarse averages of the true underlying local processes.

In contrast, it is possible to base an investigation on the more flexible nonstationarity assumption by adopting a local model, such as GWR, which allows relationships to vary by location. Only then is it possible to overcome the limitations of the stationarity assumption present in global models and avoid the pitfalls of this type of model misspecification.

2.2.1 Model Specification

As an extension of the linear regression framework, GWR can be written in terms of the notation of equation (2.1) and is specified as follows:

$$y_i = \beta_{0i} + \beta_{1i}x_{1i} + \beta_{2i}x_{2i} + \ldots + \beta_{ki}x_{ki} + \varepsilon_i \tag{2.3}$$

where the primary difference is that now parameters $\{\beta_{0i}, \beta_{1i}, \beta_{2i}, \ldots, \beta_{ki}\}$ are also indexed by i because a separate relationship is measured at each location (Fotheringham et al., 1998; Brunsdon et al., 1996). This means that the sign and magnitude of each parameter estimate represent the relationship between an explanatory variable and the dependent variable *for only a single location* in the study area. By estimating these location-specific parameters during model calibration, it becomes possible to interpret and visualize how processes manifest uniquely in different places.

2.2.2 Model Calibration

In order to calibrate the model in equation (2.3), it is necessary to estimate parameters for each location i, exposing a practical inconvenience: we typically only have a few observations or even a single observation at each location. If we wanted to use the OLS estimator in equation (2.2), we would not be able to reliably estimate the location-specific parameters for the GWR model specified in equation (2.3). To overcome this limitation, we borrow data from *nearby* locations. In GWR, this is done through the specification of an $m \times m$ spatial weights matrix $\boldsymbol{W}$ that encodes a data-borrowing scheme designed to allow data points closer to location i to have a stronger influence on the parameter estimate associated with it. Each row of $\boldsymbol{W}$ contains values between 0 and 1 that weights the importance of each observation based on its proximity to a calibration location i and can be restructured as an $m \times m$

diagonal matrix $\boldsymbol{W}_i$ with off-diagonal elements set to zero. Parameter estimates can then be obtained at each location i using a weighted version of the OLS estimator from equation (2.2)

$$\widehat{\beta}_i = [\boldsymbol{X}^T\boldsymbol{W}_i\boldsymbol{X}]^{-1}\boldsymbol{X}^T\boldsymbol{W}_i\boldsymbol{y} \tag{2.4}$$

which has the effect of shrinking the observations that are relatively far away from location i toward zero so that they have a diminishing influence on the regression. Equation (2.4) is calculated for each location i, creating an ensemble of locally weighted regressions that allow a GWR model of the form shown in equation (2.3) to be calibrated (Fotheringham et al., 2002). The selection of the spatial weighting matrix $\boldsymbol{W}$ or "data-borrowing scheme", which is characterized by a kernel function, a measure of proximity, and a bandwidth parameter that controls the intensity of weighting, is therefore a central step in any GWR analysis.

It follows then that the predicted value of each observation is given by

$$\widehat{y}_i = \boldsymbol{X}_i[\boldsymbol{X}^T\boldsymbol{W}_i\boldsymbol{X}]^{-1}\boldsymbol{X}^T\boldsymbol{W}_i\boldsymbol{y} \tag{2.5}$$

which is equivalent to multiplying the estimated parameters $\widehat{\beta}_i$ for each location by $\boldsymbol{X}_i$, a vector of the values for the observations of the explanatory variables at location i. This can also be rewritten for the entire vector of predicted values as

$$\widehat{\boldsymbol{y}} = \boldsymbol{S}\boldsymbol{y} \tag{2.6}$$

if $\boldsymbol{S}$ is constructed such that each row is calculated as

$$\boldsymbol{S}_i = \boldsymbol{X}_i[\boldsymbol{X}^T\boldsymbol{W}_i\boldsymbol{X}]^{-1}\boldsymbol{X}^T\boldsymbol{W}_i \tag{2.7}$$

The matrix $\boldsymbol{S}$ is also known as the hat matrix or the projection matrix because it can be seen in equation (2.6) that multiplying it by $\boldsymbol{y}$ projects the hat onto $\widehat{\boldsymbol{y}}$ to produce the predicted values of the dependent variable. The local scaled variance-covariance matrix of the parameter estimates, $\boldsymbol{V}_i$, can similarly be constructed in a row-wise fashion such that

$$\boldsymbol{C}_i = [\boldsymbol{X}^T\boldsymbol{W}_i\boldsymbol{X}]^{-1}\boldsymbol{X}^T\boldsymbol{W}_i \tag{2.8}$$

$$\boldsymbol{V}_i = \boldsymbol{C}_i\boldsymbol{C}_i^T\widehat{\sigma}^2 \tag{2.9}$$

where $\widehat{\sigma}^2$ is the estimated standard deviation of the error term for the model, which is defined as

$$\widehat{\sigma}^2 = \frac{\sum_i \left(y_i - \widehat{y}_i\right)^2}{m - tr\left(\boldsymbol{S}\right)} \tag{2.10}$$

where the numerator is the residual sum of squares, m is number of observations in the sample, and tr is the trace operator. The trace of the hat matrix, $tr(\boldsymbol{S})$, yields an important value called the effective number of parameters (*ENP*), which serves as a proxy for the number of degrees of freedom that are consumed by the model (Fotheringham et al., 2002).

Finally, the standard errors for the estimated parameters are calculated as

$$se\left(\widehat{\beta}_i\right) = \sqrt{diag(V_i)} \tag{2.11}$$

Examining the above equations and how many of them are indexed by i, it becomes clear that a large portion of the GWR estimation routine involves executing calculations separately for each location, a feature that allows the possibility for the routine to be parallelized and scale to accommodate models hundreds of thousands or even millions of observations (Li et al., 2019). A computation-efficient GWR estimation routine is summarized in Algorithm 2.1, which generally involves first specifying a data-borrowing scheme based on a bandwidth parameter.

2.2.3 Data-Borrowing Schemes

In order to construct $\boldsymbol{W}$ and compute $\widehat{\beta}_i$ using equation (2.4), it is necessary to specify a data-borrowing scheme, which includes selecting a kernel function, a measure of proximity (or kernel type), and a bandwidth parameter. These are each discussed in the subsequent sections in turn.

2.2.3.1 Kernel Functions

A kernel function is applied to the distances between locations for observations and calibration points in order to calculate the weights matrix. In theory, this could be any

ALGORITHM 2.1

Estimation of Geographically Weighted Regression.

1. Initialize empty $\widehat{\beta}(m \times k)$, $\hat{y}(m \times 1)$, $\boldsymbol{S}$ $(m \times m)$, $\boldsymbol{CCT}$ $(m \times k)$, $\boldsymbol{SE}$ $(m \times k)$
2. Given a data-borrowing scheme, compute $\boldsymbol{W}$
3. For location $1 \ldots m$, at each location i, calculate:
4. The diagonal matrix $\boldsymbol{W}_i$ based a row of the spatial weight matrix $\boldsymbol{W}$
5. Parameter estimates $\widehat{\beta}_i$ from equation (2.4); store in $\widehat{\beta}$
6. Predictive value $\hat{y}_i$ from equation (2.5); store in $\hat{y}$
7. Row of hat matrix $\boldsymbol{S}_i$ from equation (2.7); store in $\boldsymbol{S}$
8. Diagonal of intermediate values $\boldsymbol{C}_i\boldsymbol{C}_i^T$ from equation (2.8); store in $\boldsymbol{CCT}$
9. End for
10. Calculate $\widehat{\sigma}^2$ from equation (2.10)
11. Calculate standard errors $se\left(\widehat{\beta}_i\right)$ in equation (2.11); store in $\boldsymbol{SE}$
12. End for
13. End

function that places more emphasis on observations that are closer to calibration points than to those that are further away. However, the three most widely used kernel functions in practice are the Gaussian, exponential, and bi-square functions, which are specified in Table 2.1. Each of these three functions has a slightly different shape, which can be seen in Figure 2.1. Importantly, one key difference is that the Gaussian and exponential kernel functions allow all observations to retain nonzero (albeit potentially very small) weights regardless of how far they are from the calibration location. The result is that even far-away observations can remain influential, especially for moderate-to-large bandwidth parameters. In contrast, a bi-square kernel does not suffer from this issue because the bandwidth parameter is the exact distance, or number of nearest neighbors, away in space that all remaining observations are weighted to zero and have no influence. This is illustrated in Figure 2.1 on the left diagram where it can be seen that for the bi-square kernel some observations toward the tails of the function are weighted to exactly zero, even for larger bandwidths. There are also other kernel functions found in the literature,

TABLE 2.1

Common Kernel Function Specifications.

Function	Specification
Gaussian	$w_{ij} = exp\left(-\frac{1}{2}\left(\frac{d_{ij}}{bw}\right)^2\right)$
Exponential	$w_{ij} = exp\left(-\left(\frac{d_{ij}}{bw}\right)\right)$
Bi-square	$w_{ij} = \{\left(1-\left(\frac{d_{ij}}{bw}\right)^2\right), d_{ij} < bw 0, otherwise$

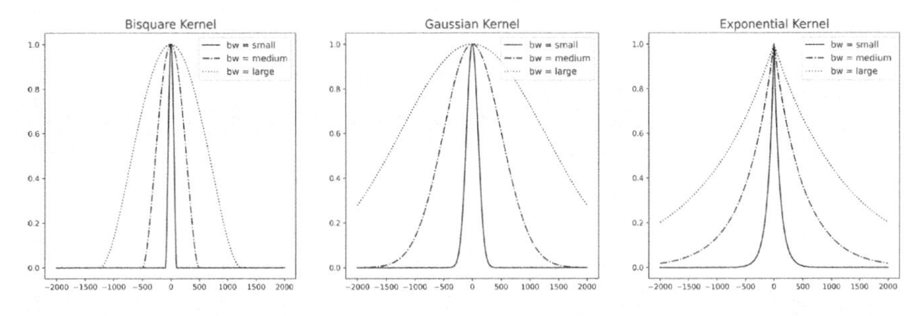

FIGURE 2.1
Examples of Bi-square Kernels (left), Gaussian Kernels (middle), and Exponential Kernels (right) for a Small, Medium, and Large Bandwidth Parameter.

Source: Originally appears in Oshan et al. (2019a).

such as the box-car function and the tri-cube function, though they are much less common in practice and are therefore not discussed here.

2.2.3.2 Kernel Types

There are two primary measures of proximity that can be considered when selecting a kernel. The first, a distanced-based kernel, is characterized by a measure of distance from the calibration location, which is *fixed* for each calibration location regardless of how many sample observations are found within this threshold. This can be a limitation when the spatial distribution of observations is heterogeneous across a study area, and some parts are more sparsely populated with data than others because under such a scenario some of the local regressions may be based on relatively small sample sizes. The second type, a nearest-neighbor kernel, avoids this issue. By using a nearest-neighbor definition of bandwidth, the width of the kernel (i.e., distance) *adapts* from location to location, ensuring that the same number of observations is available for each local regression. The difference between these two kernel types is illustrated in Figure 2.2. The fixed kernels (upper) are the same regardless of the distribution of the data while the adaptive kernels (lower) vary in shape depending upon the spatial distribution of the data. As a result, an adaptive bandwidth kernel is able to better handle irregularly shaped study areas, non-uniform spatial distributions of observations, and edge effects.

2.2.3.3 Bandwidth Selection

To use GWR as a spatial microscope, the bandwidth parameter (either distance or the number of nearest neighbors) can be set on the basis of arbitrary intervals to explore the sensitivity of the results. The bandwidth can also be set on the basis of prior belief about the range or scale at which associations are thought to exist. It often is not clear though what this scale should be in practice so the prevailing strategy is to find an optimal bandwidth by using a computational routine that optimizes a model fit diagnostic, such as minimizing a leave-one-out cross-validation criterion (Bowman, 1984) defined as

$$CV(bw) = \sum_i \left[y_i - \hat{y}_{\neq i}(bw) \right]^2 \tag{2.12}$$

where y_i is the observed value of the dependent variable at location i, and $\hat{y}_{\neq i}(bw)$ denotes the predicted value at location i for a model that leaves out the observed value at location i and is calibrated using bandwidth bw. This particular criterion only optimizes the predictive accuracy of the model, but it is also possible to consider both the model predictive accuracy and complexity (i.e., degrees of freedom consumed on the basis of the number of parameters being estimated) by using an information criterion, such as the Bayesian information criterion (*BIC*) or the Akaike information criterion (*AIC*) (Akaike, 1974; Hurvich et al., 1998). The latter of these two information criteria has been extended specifically for the case of GWR to avoid the potential overfitting that can occur for small bandwidth values (i.e., very local) associated with excessive complexity and small sample size (Fotheringham et al.,

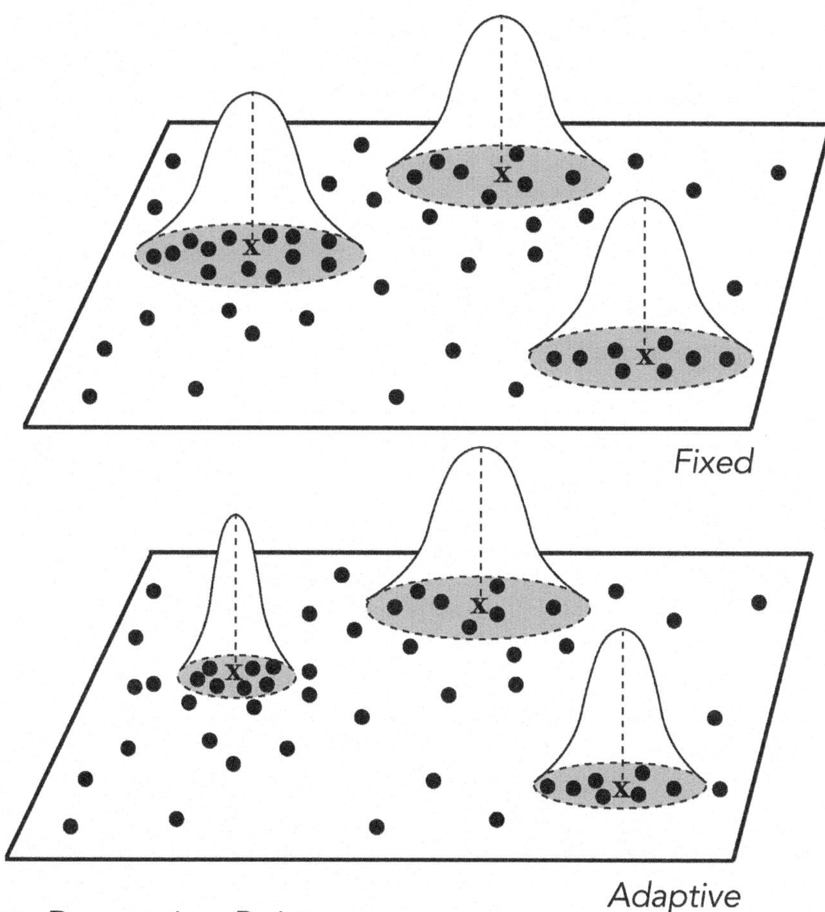

FIGURE 2.2
Fixed Versus Adaptive Kernel Function Operation.

2002). This GWR-specific *AIC* therefore incorporates a larger penalty for smaller bandwidth values to avoid this issue and is defined as

$$AICc(bw) = 2m\,ln(\hat{\sigma}) + m\,ln(2\pi) + m\left\{\frac{m + tr(\boldsymbol{S})}{m - 2 - tr(\boldsymbol{S})}\right\} \tag{2.13}$$

where *ln* denotes the natural logarithm, $\hat{\sigma}$ is the estimated standard deviation of the error term, *m* is the overall sample size, and $tr(\boldsymbol{S})$ is the trace of the hat matrix $\boldsymbol{S}$. The overall result is that a bandwidth parameter selected by minimizing this *AICc* optimizes a trade-off between bias and variance in the local parameter estimates (i.e., $\widehat{\beta}_{ki}$) based on the available data while also considering sample size (i.e., m) and model complexity (i.e., $tr(\boldsymbol{S})$) (Yu et al., 2020a). Since the bandwidth parameter controls the intensity of weighting or data borrowing, selecting it based on the $AICc$ yields the range (i.e., scale) that conditional spatial associations can be most robustly captured.

Another important consideration for bandwidth selection is how much of the parameter space to evaluate when searching for an optimal value. It is usually computationally prohibitive to exhaustively search all potential values for the bandwidth parameter, but there is also a risk of identifying a local optimum rather than the global optimum if the parameter space is not sufficiently explored. It is also possible to use an equal interval search technique that evaluates values at set intervals between a specified minimum and maximum value. This is particularly useful when trying to avoid bandwidth values above or below a given threshold or when it is desired to guarantee that a portion of the parameter space is exhaustively searched, but it requires an exogenously defined interval that is still prone to issues of computational complexity and local optima. A more efficient technique that has also become the standard algorithm implemented in most GWR software is the golden-section search strategy, which adaptively searches the parameter space to iteratively hone in on an optimal parameter value without exhaustively searching the parameter space.

2.2.4 Model Diagnostics

2.2.4.1 Local Hypothesis Testing

As an extension of the traditional regression framework, it is possible to carry out a pseudo t-test for each local parameter estimate $\widehat{\beta}_{ki}$ to test whether or not it is significantly different from the null hypothesis that $\beta_{ki} = 0$. The local t-values are calculated as

$$t_{ki} = \frac{\widehat{\beta}_{ki}}{se_{ki}} \tag{2.14}$$

where $\widehat{\beta}_{ki}$ and se_{ki} are the parameter estimate and standard error at location i for explanatory variable k, respectively. Traditionally, in a global regression model fitted with relatively large datasets, a t-value larger than $\pm$ 1.96 indicates that an estimate is different from zero at the 95% confidence level $(1-\alpha;\ \alpha = 0.05)$, suggesting that we may reject the null hypothesis in favor of the alternative hypothesis that $\beta_{ki} = \widehat{\beta}_{ki}$. However, there are two issues that need to be considered when extending this logic to the GWR framework. The first consideration is that in GWR there will be m parameter estimates for explanatory variable k that we are simultaneously testing for significance using the same sample of observations. Anytime multiple hypothesis tests are carried out on the same set of observations this is an issue because the probability of at least one false positive (incorrectly rejecting the null hypothesis in favor of the alternative hypothesis) test is $1-(1-\alpha)^m$ for m tests and a type I error rate of α. Figure 2.3 demonstrates how this probability rapidly increases toward 1 for even a modest number of tests $(m < 50)$ at the 95% confidence level $(\alpha = 0.05)$. This means that a more conservative (i.e., smaller) α-value needs to be applied to maintain the intended 5% chance of a false positive test (i.e., 95% confidence interval). The second consideration is that in GWR these multiple tests are not independent of each other because each local subset of data overlaps with neighboring subsets. This means that the number of *independent* tests, m^*, is $1 < m^* < m$, and conventional corrections for multiple testing, such as a Bonferroni correction (Bonferroni, 1935), are not appropriate for GWR.

Instead, to deal with these concerns, it is necessary to account for multiple dependent hypothesis tests through a GWR-specific correction developed by da Silva and Fotheringham (2016) that can be applied to obtain the adjusted α-value

$$\alpha^* = \frac{\xi}{\frac{ENP}{k}} \tag{2.15}$$

where ξ is the desired joint type I error rate (i.e., 0.05), ENP is the effective number of parameters in GWR that depends upon the bandwidth (i.e., data-borrowing scheme), and k is the number of explanatory variables in a given model. The ratio $\frac{ENP}{k}$ ($ENP > k$) is representative of the number of multiple tests for a given data-borrowing scheme. If $ENP = k$ then $\xi = \alpha$ and the tests performed by GWR and a global regression are equivalent. Using equation (2.15) to obtain an adjusted α^*, it is possible

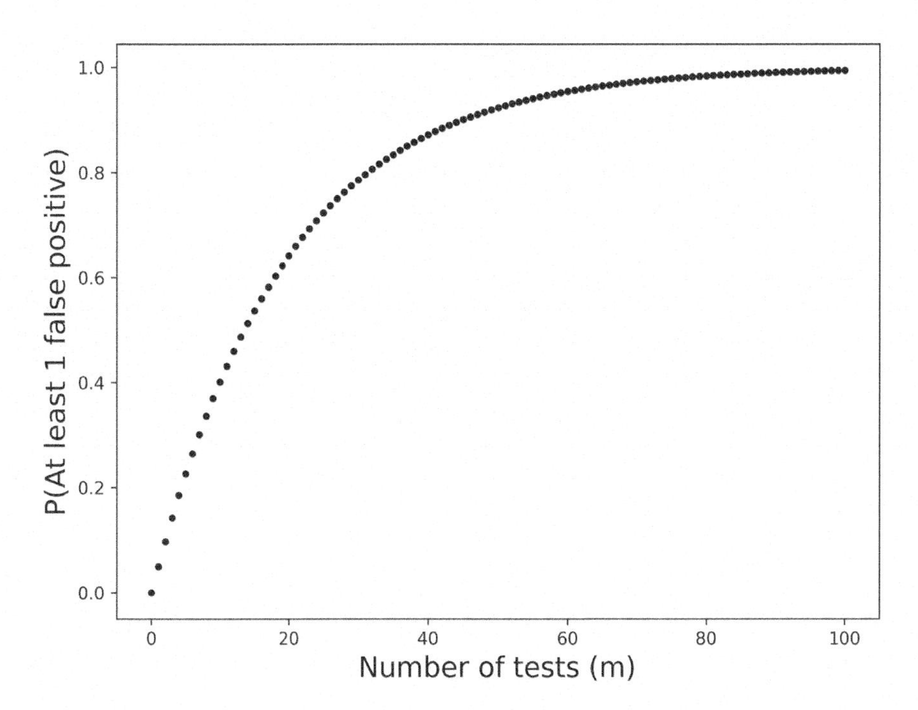

FIGURE 2.3
The Probability of at Least One False Positive Test if Conducting m Tests ($\alpha = 0.05$).

to derive a corrected critical t-value that is likely larger than ± 1.96 and is more conservative as a result. Many previous GWR applications have not employed this hypothesis testing framework, which means inference from these studies is questionable. It is therefore critical to ensure that this GWR-specific correction is applied for hypothesis testing and that these adjusted values are considered and mapped when interpreting the output of a GWR model.

2.2.4.2 Model Fit

In most regression exercises it is of interest to gauge the quality of how a model fits the data being modeled, and this is also true for GWR. The traditional *AIC* or the *AICc* denoted in equation (2.13) can be used to compare the fit of alternative models with lower values indicating better model fit. It is also possible to use an R^2 value to gauge the amount of variation being accounted for using a ratio of the residual sum of squares over the total sum of squares such that

$$R^2 = 1 - \frac{RSS}{TSS} \tag{2.16}$$

where

$$RSS = \sum_i \left(y_i - \hat{y}_i\right)^2 \tag{2.17}$$

and

$$TSS = \sum_i \left(y_i - \bar{y}_i\right)^2 \tag{2.18}$$

An advantage of the GWR framework is that we can also explore model fit for each calibration location using a local version of the R^2

$$R_i^2 = 1 - \frac{RSS_i}{TSS_i} \tag{2.19}$$

where

$$RSS_i = \sum_i w_i \left(y_i - \hat{y}_i\right)^2 \tag{2.20}$$

and

$$TSS_i = \sum_i w_i \left(y_i - \bar{y}_i\right)^2 \tag{2.21}$$

where $\boldsymbol{w}$ is the $m \times 1$ vector of spatial weights for location i (as opposed to the diagonal matrix $\boldsymbol{W}_i$). Since this model fit criterion is available for each location i, it can be mapped in order to visually explore where a model fit is particularly strong or weak.

2.2.4.3 Testing for Spatial Variability

It is also possible to test the statistical significance of spatial patterns for each surface of parameter estimates produced by GWR via Monte Carlo methods. The *spatial variability test* shuffles the observations in space, re-calibrates GWR on the randomized data while holding the model specification constant, and then computes the variability of the resulting parameter estimates for each surface. This process is repeated, and the number of times that the variability of each surface from the randomized data is higher than the variability of each original surface is used to construct pseudo p-values for hypothesis testing. A pseudo p-value smaller than 0.05 indicates that the observed spatial variability of a coefficient surface is significant at the 95% confidence level so that the null hypothesis that the original surface is globally constant can be rejected in favor of the alternative hypothesis that the original surface is not constant. A pseudo p-value larger than .05 indicates that the null hypothesis cannot be rejected, suggesting that the spatial variation in the original surface is not significantly different from that expected due to noise. One issue with the test for spatial variability is that it requires GWR to be calibrated many times, which can become computationally prohibitive for larger datasets. Users should exercise caution in how many replications they specify for the test, keeping in mind that the suggested number of iterations is 1,000.

2.2.4.4 Detecting Multicollinearity

Traditional tests for detecting multicollinearity can also be localized in the case of GWR to explore "pockets" of collinearity in the weighted subsets of data that could affect the local parameter estimates in a similar manner to that of the global parameter estimates from a traditional regression model. These diagnostics are useful because it is possible to have higher levels of collinearity among each subset than is in the overall unweighted data (Wheeler & Tiefelsdorf, 2005). In general, higher levels of collinearity are associated with problems such as parameter estimate instability, unintuitive parameter signs, high R^2 diagnostics despite few or no significant parameters, and inflated standard errors for parameter estimates (Belsey et al., 1980; O'Brien, 2007). Diagnostic tools for detecting levels of *local* multicollinearity include local correlation coefficients (*CC*), local variation inflation factor (*VIF*), the local condition number (*CN*), and local variance decomposition proportions (*VDP*) (Wheeler & Tiefelsdorf, 2005; Wheeler, 2007), which are each subsequently described in more detail.

Local correlation coefficients provide an extension of the traditional Pearson's correlation coefficient to investigate collinearity between pairs of variables. For each pair of variables (k,l), the local correlation coefficients are specified as

$$r_i(k,l) = \frac{\sum_{j=1}^{m} w_{ij}^{*}\left(x_{kj} - \bar{x}_{ki}\right)\left(x_{lj} - \bar{x}_{li}\right)}{\left[\sum_{j=1}^{m} w_{ij}^{*}\left(x_{kj} - \bar{x}_{ki}\right)\sum_{j=1}^{m}\left(x_{lj} - \bar{x}_{li}\right)\right]} \quad (2.22)$$

where x is the value of variables k or l at locations i or j and where

$$\bar{x}_{li} = \sum_{j=1}^{m} w_{ij}^{*} x_{lj} \tag{2.23}$$

is the weighted mean of explanatory variable l at location i (or variable k at location i), and w_{ij}^{*} is the square root of the kernel weight between locations i and j standardized such that they sum to 1. Similar to traditional correlation coefficients, these local correlation coefficients also range between values of -1 to 0 for an inverse relationship and 0 to 1 for a direct relationship. A useful rule of thumb for correlation coefficients is that values greater than $|0.8|$ may indicate potential collinearity issues within a regression, though this will ultimately depend upon the number of variables under consideration, the correlation between the other pairs of variables, and the sample size (Páez et al., 2011). The limitation of correlation coefficients is that they do not shed light on the collective level of multicollinearity among a group of variables nor do they provide information about which individual variable may be contributing the most toward potential issues.

A local variance inflation factor (*VIF*) can provide information about the individual contribution of variables toward multicollinearity issues and can be obtained for each calibration location as

$$VIF_i(k) = \frac{1}{1 - R_i^2(k)} \tag{2.24}$$

where $R_i^2(k)$ is the coefficient of determination at location i when explanatory variable x_k is regressed with geographical weighting using the GWR optimized bandwidth on the other variables for the same location. One rule of thumb is that to avoid severe issues associated with multicollinearity, the *VIF* for each variable should be less than 10. However, a more conservative approach is to filter out variables with a *VIF* higher than 5 in order to avoid more mild issues associated with multicollinearity or even to use a rule of thumb of 3 to ensure that there are no potential issues. One drawback of the *VIF* diagnostic is that it does not consider potential correlation between variables and the intercept.

The local condition number provides another diagnostic that does consider potential correlation with the intercept and allows collinearity to be assessed collectively among a group of variables. It can be obtained through a singular value decomposition (*SVD*) of each local subset of data. Formally, this is denoted as

$$\sqrt{\boldsymbol{W}_i}\boldsymbol{X} = \boldsymbol{U}\boldsymbol{D}\boldsymbol{V}^T \tag{2.25}$$

where $\boldsymbol{U}$ and $\boldsymbol{V}$ are orthogonal $m \times (k+1)$ and $(k+1) \times (k+1)$ matrices, respectively, $\boldsymbol{D}$ is a $(k+1) \times (k+1)$ matrix containing a diagonal of decreasing singular values, $\sqrt{\boldsymbol{W}_i}$ is the square root of the diagonal weight matrix for location i, and $\mathbf{X}$ is the column norm-scaled $m \times k$ matrix of explanatory variables inclusive of the intercept. The condition number for location i can be calculated using the singular values from $\boldsymbol{D}$ associated with $\sqrt{\boldsymbol{W}_i}\boldsymbol{X}$ such that

$$CN = \frac{D_{max}}{D_{min}} \quad (2.26)$$

where D_{max} and D_{min} are the largest (first) and smallest (last) singular value along the diagonal of $\boldsymbol{D}$, respectively. Condition numbers larger than 30 are often used to denote severe issues associated with multicollinearity, though a more conservative rule of thumb that is often employed is to use a condition number of 10 or below to denote scenarios where multicollinearity issues are not likely.

Using the *SVD* of $\sqrt{\boldsymbol{W}_i}\boldsymbol{X}$, it is also possible to obtain a $l \times k$ $(l = k)$ matrix of variance decomposition proportions for each location i to further explore potential variable-specific contributions to multicollinearity issues

$$\pi_{lk} = \frac{\phi_{kl}}{\phi_k} \quad (2.27)$$

where

$$\phi_{kl} = \frac{v_{kl}^2}{d_l^2} \quad (2.28)$$

and

$$\phi_k = \sum_l \phi_{kl} \quad (2.29)$$

where v_{kl}^2 are the elements of the $\boldsymbol{V}$ matrix and d_l^2 are the singular values from the diagonal of $\boldsymbol{D}$. Rather than reporting and interpreting the entire $l \times k$ matrix π_{lk}, it has become standard practice to use the lth row of π_{lk} to explore the impact of collinearity on each $\widehat{\beta}_k$ associated with explanatory variable k. It has been suggested that if two or more elements of the lth row of π_{lk} are relatively large (greater than 0.5), it may indicate that there are issues associated with multicollinearity (Belsey et al., 1980).

Importantly, the rules of thumb described here are not absolute, and values lower than these thresholds do not necessarily guarantee that multicollinearity is not problematic nor that values above these thresholds cannot produce meaningful regression results (O'Brien, 2007; Fotheringham & Oshan, 2016). Due to this, and that each diagnostic has some advantages over the others, it has become common to rely upon several of these diagnostics collectively to inform about any potential issues rather than any one in particular. Further comments on the issue of multicollinearity in local models can be found in Section 6.3.3.

2.3 Multiscale Geographically Weighted Regression

A major limitation of the basic GWR framework described earlier is that the same bandwidth is assumed to apply for each relationship in the model, which means the data are weighted at the same *spatial scale*. This limitation is overcome by MGWR (Fotheringham et al., 2017), which relaxes this assumption and instead allows a potentially unique bandwidth to be specified for each relationship in the model. The advantage of the MGWR extension is that it can more accurately capture the spatial heterogeneity within and across spatial processes, minimize overfitting, mitigate concurvity (i.e., collinearity due to similar functional transformations), and reduce bias in the parameter estimates (Fotheringham et al., 2017; Wolf et al., 2018; Oshan et al., 2019b; Yu et al., 2020a).

2.3.1 Model Specification

The assumption that the same bandwidth is used for every relationship in the model can be relaxed to extend the GWR specification in equation (2.3), such that

$$y_i = bw_0\left(\beta_{0i}\right) + bw_1\left(\beta_{1i}x_{1i}\right) + bw_2\left(\beta_{2i}x_{2i}\right) + \ldots + bw_k\left(\beta_{ki}x_{ki}\right) + \varepsilon_i \tag{2.30}$$

where $bw_0, bw_1, bw_2, \ldots, bw_k$ are the individual bandwidths used to parameterize a separate data-borrowing scheme that is applied to each model component. In theory, it is possible to specify different data-borrowing schemes for different model components, such as an adaptive bi-square spatial kernel for $bw_0\left(\beta_{0i}\right)$, a time-based kernel for $bw_1\left(\beta_{1i}x_{1i}\right)$, and a network connectivity-based kernel for $bw_2\left(\beta_{2i}x_{2i}\right)$ (Lu et al., 2014). However, this makes it difficult to compare the bandwidths to each other in a meaningful way, since each would be associated with a different measurement unit. Furthermore, non-spatial kernels are not straightforward to interpret in terms of geographical context. The ability to comparatively interpret the bandwidths as an indicator of the spatial scales at which processes operate is a major advantage of MGWR over GWR, and it has therefore become standard practice to employ the same spatial kernel for each model component. After selecting a spatial kernel, the next step for specifying an MGWR model is to identify appropriate values for the individual bandwidths, which, like GWR, is most commonly done computationally as part of the model calibration routine.

2.3.2 Model Calibration via Backfitting

MGWR model calibration accommodates the selection and incorporation of relationship-specific bandwidths (i.e., multiple scales) by reformulating GWR as a generalized additive model (GAM)

$$y = f_{bw_0} + f_{bw_1}\left(X_1\right) + f_{bw_2}\left(X_2\right) + \ldots + f_{bw_k}\left(X_k\right) + \varepsilon \tag{2.31}$$

where $f_{bw_k}(X_k) = bw_k(\beta_k)X_k$ is a smoothing function (i.e., the data-borrowing scheme) applied to the kth explanatory variable X_k. To estimate each smoothing function f_{bw_k} and calculate parameters β_k, equation (2.31) can be rearranged, and knowing ε is independent to X_k gives the following component-wise conditional expectation (Li & Fotheringham, 2020)

$$f_{bw_k} = E\left(y - \sum_{p \neq k} f_{bw_p} - \varepsilon \mid X_k\right) = E\left(y - \sum_{p \neq k} f_{bw_p} \mid X_k\right) = A_k\left(y - \sum_{p \neq k} f_{bw_p}\right) \quad (2.32)$$

where A_k is $E(\cdot \mid X_k)$, which is the hat matrix from a univariate GWR model of X_k that maps $y - \sum_{p \neq k} f_{bw_p}$ to $\hat{f}_{bw_k}$. This GWR hat matrix A_k is expressed as

$$A_k = \begin{pmatrix} x_{1k}\left(X_k^T W_1 X_k\right)^{-1} X_k^T W_1 \\ \cdots \\ x_{mk}\left(X_k^T W_m X_k\right)^{-1} X_k^T W_m \end{pmatrix}_{mxm} \quad (2.33)$$

where W_i is a diagonal spatial weight matrix calculated on the basis of a covariate-specific bandwidth and a kernel function (e.g., bi-square or Gaussian). Putting all the additive components in equation (2.32) into a matrix form gives a normal system of

$$\begin{bmatrix} I & A_1 & \cdots & A_1 \\ A_2 & I & \cdots & A_2 \\ \vdots & \vdots & \ddots & \vdots \\ A_k & A_k & \cdots & I \end{bmatrix} \begin{bmatrix} f_{bw_1} \\ f_{bw_2} \\ \vdots \\ f_{bw_k} \end{bmatrix} = \begin{bmatrix} A_1 \\ A_2 \\ \vdots \\ A_k \end{bmatrix} y \quad (2.34)$$

Equation (2.34) can be written in abbreviated form as

$$Pf = Qy \quad (2.35)$$

where P is an mk-by-mk matrix and Q is an mk by m matrix. On rearranging, this gives additive components $f_{bw_1}, \cdots f_{bw_k}$ as

$$f = \begin{bmatrix} f_{bw_1} \\ f_{bw_2} \\ \vdots \\ f_{bw_k} \end{bmatrix} = P^{-1}Qy \quad (2.36)$$

provided that matrix $\boldsymbol{P}$ is invertible. Then, covariate-specific hat matrices can be obtained by

$$\boldsymbol{R} = \begin{bmatrix} \boldsymbol{R}_1 \\ \boldsymbol{R}_2 \\ \vdots \\ \boldsymbol{R}_k \end{bmatrix} = \boldsymbol{P}^{-1}\boldsymbol{Q} \tag{2.37}$$

This normal system of equations (2.35) provides a closed form of MGWR parameter estimates and inference. From a computation perspective, directly solving equation (2.35) is often prohibitive for even moderately sized datasets. The solution to this is to use an iterative backfitting procedure originally formulated by Hastie and Tibshirani (1986) and Buja et al. (1989), which converges toward a stable solution as directly solved in equation (2.35). The backfitting routine also derives a set of data-driven bandwidth parameters for the k processes being modeled and is optimal in the sense that they minimize the overall model fit while balancing a bias-variance trade-off for each individual model component. This means that each model component comprises a univariate GWR model, and each iteration of the backfitting procedure entails a GWR calibration based on a subset of the steps outlined in Algorithm 2.1.

The MGWR backfitting calibration, which is outlined in Algorithm 2.2, begins by first setting an initial value for the local parameter estimates $\widehat{\beta}_{i_{MGWR}}$, which could be zero ($\widehat{\beta}_{i_{MGWR}} = 0 \forall i$), the global parameter estimates from a traditional regression ($\widehat{\beta}_{i_{MGWR}} = \widehat{\beta} \, \forall i$), or the local parameter estimates from a GWR model including all of the variables ($\widehat{\beta}_{i_{MGWR}} = \widehat{\beta}_{i_{GWR}}$). The closer the initial local parameter estimates are to the eventual converged values, the fewer the iterations that are needed and the quicker the algorithm can be executed. Fotheringham et al. (2017) demonstrate that using initial parameter estimate values from GWR is typically superior in this regard.

These initial values for the local parameter estimates are then used to obtain initial values for predicted values $\hat{y}$ and residuals $\hat{\varepsilon}$. During each iteration of the backfilling procedure, for each model term k, the current value of $\widehat{\boldsymbol{f}_k} + \hat{\varepsilon}$ is regressed on $\boldsymbol{X}_k$ using GWR, providing a temporary value for the optimal bandwidth bw_k associated with the relationship between $\boldsymbol{y}$ and $\boldsymbol{X}_k$ and updated values for the local parameter estimates $\widehat{\beta}_k$, smooth $\widehat{\boldsymbol{f}_k}$, and model residuals $\hat{\varepsilon}$ (based on these other updated values). At the end of each iteration, the current values of $\hat{\boldsymbol{y}}$ and $\hat{\varepsilon}$ are used to assess whether the differences between these values and the previous iteration are sufficiently small to denote that the backfilling algorithm has converged. A score of change (SOC) criterion can be used to assess convergence, which can be based on the MGWR model RSS

$$SOC_{RSS} = \frac{RSS_{new} - RSS_{old}}{RSS_{new}} \tag{2.38}$$

or based directly on the individual GWR-style smoothing functions

$$SOC_{sf} = \sqrt{\frac{\sum_k \frac{\sum_i (\hat{f}_{ki}^{new} - \hat{f}_{ki}^{old})^2}{n}}{\sum_i \left(\sum_k \hat{f}_{ki}^{new}\right)^2}} \tag{2.39}$$

Though both SOC_{RSS} and SOC_{sf} are scale-free, the SOC_{sf} is advantageous because it focuses on the changes of each model component rather than the overall model fit and is therefore the suggested criterion to use. Convergence is achieved once the SOC becomes smaller than some threshold value η, which is typically set to 10^{-5}.

Following Yu et al. (2020b), it is also possible to compute a covariate-specific hat matrix $\boldsymbol{R}_k$ that maps the dependent variable $\boldsymbol{y}$ to each of the estimated model components $\hat{\boldsymbol{f}}_k$ in the backfitting routine such that

$$\hat{f}_k = \boldsymbol{R}_k \boldsymbol{y} \tag{2.40}$$

where $\boldsymbol{R}_k$ is computed as

$$\boldsymbol{R}_k = \boldsymbol{A}_k \left(\boldsymbol{I} - \sum_{p \neq k} \boldsymbol{R}_p \right) \tag{2.41}$$

and $\boldsymbol{A}_k$ is the hat matrix from each univariate GWR model used to estimate $\hat{\boldsymbol{f}}_k$. The calculation of $\boldsymbol{R}_k$, $\boldsymbol{A}_k$, and $\boldsymbol{R}_p$ needs to be carried out and updated after each iteration of the backfilling procedure. Then, once convergence is reached, the final values of $\boldsymbol{R}_k$ can be summed to obtain the overall hat matrix $\boldsymbol{S}$ for the MGWR model

$$\boldsymbol{S} = \sum_k \boldsymbol{R}_k, \tag{2.42}$$

which maps $\boldsymbol{y}$ to $\hat{\boldsymbol{y}}$. In addition, the covariate-specific hat matrix, $\boldsymbol{R}_k$, can be used to compute the covariance-specific effective number of parameters in the following manner

$$ENP_k = tr(\boldsymbol{R}_k) \tag{2.43}$$

and the model effective number of parameters can be calculated as

$$ENP_{model} = \sum_k ENP_k \tag{2.44}$$

2.3.3 Model Diagnostics

2.3.3.1 *Covariate-Specific Local Hypothesis Testing*

Another benefit of using MGWR over GWR is that the distinct bandwidth parameters, bw_k, and effective numbers of parameters, ENP_k, allow a unique adjusted α-value and critical t-value to be computed for each of the k relationships being modeled. For the kth set of parameter estimates, hypothesis testing is carried out using

$$\alpha_k^* = \frac{\xi}{ENP_k} \tag{2.45}$$

where ENP_k is the effective number of parameters for the kth model term (Yu et al., 2020b). Each α_k^* can then be used to derive a covariate-specific critical t-value that may differ from ± 1.96 and therefore facilitates a more nuanced analysis of the heterogeneity of the inferred spatial processes, especially when visualizing maps of each relationship-specific set of parameter estimates.

2.3.3.2 *Consequences of Covariate-Specific Bandwidths*

There are several important consequences for the MGWR modeling framework that stem from having covariate-specific bandwidths. Chief among these is that in order to effectively compare the values of the estimated bandwidth to each other, it is necessary to first standardize the input data so that $\boldsymbol{y}$ and each column of $\boldsymbol{X}$ have a mean of zero and variance of one before using the data in the MGWR calibration routine. This normalizes the magnitude and dispersion of each explanatory variable so that the covariate-specific bandwidths can be interpreted relative to each other as

ALGORITHM 2.2

Estimation of Multiscale Geographically Weighted Regression.

1. Calibrate a GWR model $\boldsymbol{y} \sim \boldsymbol{X}$ and obtain initial values for each $\widehat{\boldsymbol{f}}_k$, $\hat{\varepsilon}$, and $\boldsymbol{R}_k$
2. Initialize $SOC \gg \eta$
3. Do until $SOC < \eta$:
4. For each term k:
5. Calibrate univariate GWR model $\left(\widehat{\boldsymbol{f}}_k + \hat{\varepsilon}\right) \sim \boldsymbol{X}_k$ to obtain new $\widehat{\boldsymbol{f}}_k^*$ and $\hat{\varepsilon}^*$
6. Update $\widehat{\boldsymbol{f}}_k \leftarrow \widehat{\boldsymbol{f}}_k^*$ and $\hat{\varepsilon} \leftarrow \hat{\varepsilon}^*$
7. Calculate $\boldsymbol{R}_k$
8. End for
9. Calculate new SOC^* and update $SOC \leftarrow SOC^*$
10. End do
11. For each term k:
12. Calculate ENP_k
13. End for
14. Calculate $\boldsymbol{S}$

indicators of the spatial scale of the conditional relationships captured by the model (Fotheringham et al., 2017). Without this step, a bandwidth may be reflective of the spatial scale of a relationship but also could be influenced by the idiosyncrasies of the particular explanatory variable and how it is measured.

Another consequence of covariate-specific bandwidths is that certain diagnostics become unavailable because it is not straightforward to incorporate more than one spatial weight matrix into their respective calculations (Oshan et al., 2020). Two such examples include the local R_i^2 as a measure of model fit for each local regression within the MGWR ensemble and the local correlation coefficient that can be used to explore collinearity between pairs of variables for the subsets of locally weighted data. Extending these diagnostics is an avenue for future research.

References

Akaike, H. (1974). A new look at the statistical model identification. *IEEE Transactions on Automatic Control*, *19*(6), 716–723.

Belsey, D. A., Kuh, E., & Welsch, R. E. (1980). *Regression diagnostics: Identifying influential data and sources of collinearity*. New York: Wiley.

Bonferroni, C. E. (1935). Il calcolo delle assicurazioni su gruppi di teste. In *Studi in Onore del Professor Salvatore Ortu Carboni* (pp. 13–60). Rome: Tipografia des Senato del dott. G. Bardi.

Bowman, A. W. (1984). An alternative method of cross-validation for the smoothing of density estimates. *Biometrika*, *71*(2), 353–360.

Brunsdon, C., Fotheringham, A. S., & Charlton, M. E. (1996). Geographically weighted regression: A method for exploring spatial nonstationarity. *Geographical Analysis*, *28*(4), 281–298.

Buja, A., Hastie, T., & Tibshirani, R. (1989). Linear smoothers and additive models. *The Annals of Statistics*, *17*(2), 453–510.

da Silva, A. R., & Fotheringham, A. S. (2016). The multiple testing issue in geographically weighted regression. *Geographical Analysis*, *48*(3), 233–247.

Fotheringham, A. S., Brunsdon, C., & Charlton, M. E. (2002). *Geographically weighted regression: The analysis of spatially varying relationships*. London: Wiley.

Fotheringham, A. S., Charlton, M. E., & Brunsdon, C. (1998). Geographically weighted regression: A natural evolution of the expansion method for spatial data analysis. *Environment and Planning A*, *30*(11), 1905–1927.

Fotheringham, A. S., & Oshan, T. M. (2016). Geographically weighted regression and multicollinearity: Dispelling the myth. *Journal of Geographical Systems*, *18*(4), 303–329.

Fotheringham, A. S., Yang, W., & Kang, W. (2017). Multi-scale geographically weighted regression (MGWR). *Annals of the American Association of Geographers*, *107*(6), 1247–1265.

Hastie, T., & Tibshirani, R. (1986). Generalized additive models. *Statistical Science*, *1*(3), 297–310.

Hurvich, C. M., Simonoff, J. S., & Tsai, C.-L. (1998). Smoothing parameter selection in nonparametric regression using an improved Akaike information criterion. *Journal of the Royal Statistical Society: Series B (Statistical Methodology)*, *60*(2), 271–293.

Li, Z., & Fotheringham, A. S. (2020). Computational improvements to multi-scale geographically weighted regression. *International Journal of Geographical Information Science, 34*(7), 1378–1397.

Li, Z., Fotheringham, A. S., Li, W., & Oshan, T. M. (2019). Fast geographically weighted regression (FastGWR): A scalable algorithm to investigate spatial process heterogeneity in millions of observations. *International Journal of Geographical Information Science, 33*(1), 155–175.

Lu, B., Charlton, M., Harris, P., & Fotheringham, A. S. (2014). Geographically weighted regression with a non-Euclidean distance metric: A case study using hedonic house price data. *International Journal of Geographical Information Science, 28*(4), 660–681.

O'Brien, R. M. (2007). A caution regarding rules of thumb for variance inflation factors. *Quality & Quantity, 41*(5), 673–690.

Oshan, T. M., Li, Z., Kang, W., Wolf, L. J., & Fotheringham, A. S. (2019a). MGWR: A Python implementation of multiscale geographically weighted regression for investigating process spatial heterogeneity and scale. *ISPRS International Journal of Geo-Information, 8*(6), 269.

Oshan, T. M., Smith, J. P., & Fotheringham, A. S. (2020). Targeting the spatial context of obesity determinants via multiscale geographically weighted regression. *International Journal of Health Geographics, 19*(1), 11.

Oshan, T. M., Wolf, L. J., Fotheringham, A. S., Kang, W., Li, Z., & Yu, H. (2019b). A comment on geographically weighted regression with parameter-specific distance metrics. *International Journal of Geographical Information Science, 33*(7), 1289–1299.

Páez, A., Farber, S., & Wheeler, D. (2011). A simulation-based study of geographically weighted regression as a method for investigating spatially varying relationships. *Environment and Planning A, 43*(12), 2992–3010.

Wheeler, D. C. (2007). Diagnostic tools and a remedial method for collinearity in geographically weighted regression. *Environment and Planning A, 39*(10), 2464–2481.

Wheeler, D. C., & Tiefelsdorf, M. (2005). Multicollinearity and correlation among local regression coefficients in geographically weighted regression. *Journal of Geographical Systems, 7*(2), 161–187.

Wolf, L. J., Oshan, T. M., & Fotheringham, A. S. (2018). Single and multiscale models of process spatial heterogeneity. *Geographical Analysis, 50*(3), 223–246.

Yu, H., Fotheringham, A. S., Li, Z., Oshan, T. M., Kang, W., & Wolf, L. J. (2020b). Inference in multiscale geographically weighted regression. *Geographical Analysis, 52*(1), 87–106.

Yu, H., Fotheringham, A. S., Li, Z., Oshan, T. M., & Wolf, L. J. (2020a). On the measurement of bias in geographically weighted regression models. *Spatial Statistics, 38*, 100453.

3

Inference

3.1 Introduction

The concepts behind, and the essentials of, MGWR are introduced in the first two chapters. In this chapter, we focus on the inferential aspects of MGWR. Consider that the data we observe and measure at different locations result from a set of generally unseen spatial processes. Understanding how these processes operate helps us to understand why things are the way they are and how we can most effectively modify our environment to achieve our goals. One useful way of uncovering information on spatial processes is to formulate and calibrate spatial models and to obtain estimates of the model parameters. The more parameters we can reliably estimate, the more information on spatial processes we can generate. Models with relatively few parameters, such as a global linear regression model, might provide an overly simple abstraction of reality, which in turn might yield highly misleading inferences about the processes that have generated the data we are modeling. Local models such as MGWR contain many more parameters than global models and hence are better able to capture any intricacies of the processes being modeled, particularly when these processes are spatially varying. However, when we report spatially varying parameter estimates from a local model calibration how can we check that the spatial variation might not just be a result of sampling variation? That is, even if a set of processes is global, when we calibrate a local model capturing such processes, it would be almost certain that we would report some variation in the local parameter estimates because each estimate is derived from a differently weighted subset of the data. We therefore need to be able to separate variation in local parameter estimates that is due to sampling variation from variation that is sufficiently large for us to infer some other cause of the variation, such as the modeled processes being spatially nonstationary. Consequently, statistical inference in MGWR is important if we are to make plausible statements about the characteristics of the underlying processes being modeled.

Indeed, there are three main types of inferential issues that need to be addressed in MGWR (Fotheringham, 2023). The first concerns the inference about the *individual* local parameters estimates, which involves a standard t-test in classical regression. However, in the local modeling framework, inference is not being made about a single local parameter estimate but about m estimates where m is the number of

DOI: 10.1201/9781003435464-3

locations for which the local model is calibrated. Hence, the inferential test will be subject to the issue of multiple hypothesis testing, which is further complicated by the fact that the tests for estimates from MGWR will not be independent so that special adjustments are needed, although these turn out to be easy to apply.

The second concerns the inference about the *set* of local estimates. For each relationship specified in a model, MGWR will generate a set of local parameter estimates instead of just the single, average estimate obtained in the calibration of traditional global models. This set of localized parameter estimates will exhibit a certain degree of spatial variation—if we map the estimates, we see some variation in the surface. The question is: "*Is there sufficient variation in these local estimates for us to think that something else caused this other than just sampling variation?*" This can be assessed by testing the null hypothesis of spatial stationarity by generating a null distribution using Monte Carlo simulation or by using a theoretical distribution against which to test our null hypothesis of no spatial variation.

The third issue concerns the inference about the covariate-specific optimal *bandwidths*. In an MGWR framework, when bandwidths vary across the covariates, understanding bandwidth uncertainty provides a measure to differentiate local, regional, and global processes by comparing the optimal bandwidths, accounting for their uncertainties, across the different covariates. It would be naive to simply compare the covariate-specific bandwidths that are obtained on the basis of the single observed dataset and draw a conclusion that one process is more local or global than another. It is quite possible that the covariate-specific bandwidths are different by chance and subject to the sampling variation of the uncaptured noise. Understanding covariate-specific bandwidth uncertainty is thus crucial to being able to make inferences about the different spatial scales over which processes operate. The bandwidth uncertainties can be assessed using a bootstrap distribution or by computing Akaike weights.

We now elaborate on these three inferential issues and present examples of the inferential tests in subsequent sections.

3.2 Inference About *Individual* Local Parameter Estimates

3.2.1 The Problem

The inferential process for local parameter estimates obtained in the calibration of MGWR models involves conducting a classic t-test for each estimate to examine whether the relationship that the parameter estimate represents is statistically significant at each location. However, because the parameter is estimated for every location, multiple tests are made on the same hypothesis, which creates a well-known problem where the chance of observing false positives is erroneously inflated when many inferences are made simultaneously. As an intuitive example, if 1,000 tests are conducted, each having a type I error rate of 0.05, it is expected that 50 false positives would be generated creating the spurious impression that there is something interesting going on when there is not. This problem is compounded in local modeling because there can be substantial dependency between the data used at neighboring locations where the local model is calibrated so any false positives often form a spatial cluster, which adds to the impression that there is a real effect in certain parts of the map. We therefore need to amend traditional significance testing procedures to allow for multiple testing and for the dependency between the tests.

3.2.2 The Solution

The solution has two steps. First, we need to define a measure of the degree of dependency across the tests; second, we need to incorporate this measure of dependency into established solutions to the multiple hypothesis testing problem in inferential testing procedures.

Step 1: We need a measure of the degree of dependency, D, which ranges between 0 and 1, with 0 indicating no dependency between the local parameter estimates (the local parameter estimates are independent of each other) and 1 indicating total dependency (the local parameter estimates are identical, having been estimated from the same data).

Let k = the number of parameters (including the intercept) in the global model.

Let m = the number of locations at which the local model is calibrated.

Let ENP_j = the equivalent number of independent parameter estimates estimated for covariate j in the calibration of a model by MGWR.

Then a measure of the degree of dependency across the set of the local parameter estimates of covariate j can be written as

$$D_j = \frac{m - ENP_j}{m - 1} \tag{3.1}$$

As the bandwidth increases from 0 to infinity, ENP_j tends to 1, and D_j tends to 1. This is a situation where the local parameter estimates are the same everywhere and the relationship is a global one. The data used to calibrate each local regression are the same and comprise the total set of data, all of which are weighted 1. Hence the optimal bandwidth is infinity and ENP_j will equal 1 so that the numerator in equation (3.1) equals $m-1$ and D_j equals 1.

As the bandwidth decreases, ENP_j tends to m, and D_j tends to 0. This is a situation where each local regression is independent of every other one and would occur if there were sufficient data at each location to perform an independent set of regressions. There would then be m independent parameter estimates for covariate X_j in the set of local regressions so ENP_j would equal m and the numerator in equation (3.1) would equal 0, and hence D_j would equal 0. This represents a situation where there is no dependency between the local parameter estimates. The optimal bandwidth would tend to 0 as the data for each local regression are drawn entirely from the regression location.

In between these two extreme cases, there will be some degree of dependency across the local parameter estimates, and this will be reflected by the reported value of ENP_j being between 1 and m and the values of D_j being between 0 and 1. In short, as the optimal bandwidth increases, ENP_j decreases and D_j increases. The relationship between bandwidth and ENP_j and D_j is illustrated in Figure 3.1.[1]

Step 2: We need to adjust standard corrections to the classical significance testing procedure according to the degree of dependency D defined in equation (3.1). Corrections for multiple hypothesis testing fall into two broad classes: (i) familywise error rate (FWER), such as the Bonferroni (1936) and Šidák (1967) approaches; and (ii) false discovery rate (FDR), such as the Benjamini and Hochberg (1995) approach. Recognizing that if we perform m independent tests,

$$\begin{aligned} prob(\textit{at least one significant result}) &= 1 - prob(\textit{no significant results}) \\ &= 1 - (1-\alpha)^m \end{aligned} \tag{3.2}$$

where α is the type I error rate and m is the number of tests. It is easy to see how this probability increases as the number of tests m increases. FWER approaches aim to

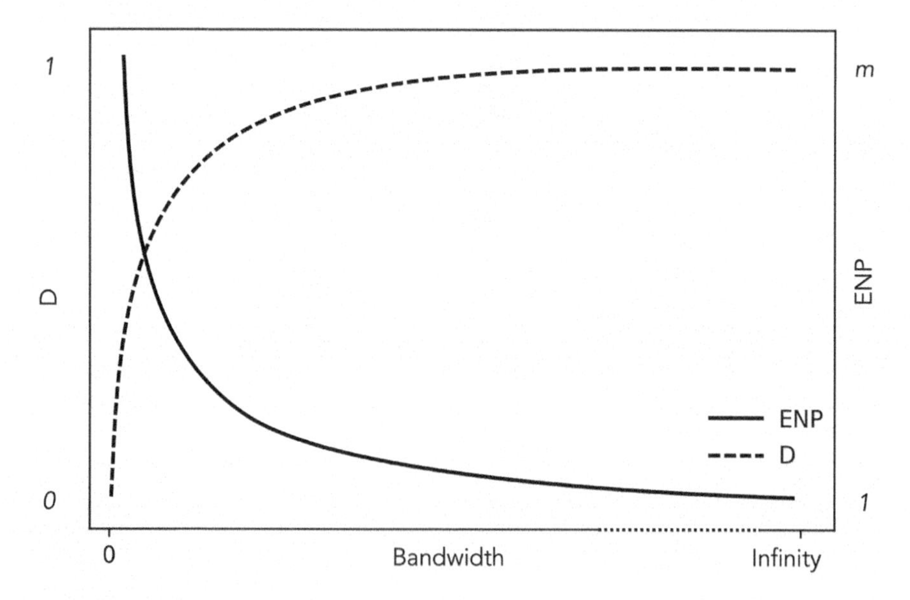

FIGURE 3.1
Relationship Between the Bandwidth, the Effective Number of Parameters (*ENP*), and the Degree of Dependency (*D*).

control the probability that at least one true null hypothesis is incorrectly rejected so that the value is close to α no matter what the value of m is.

3.2.3 Example 1: Modified Bonferroni (FWER)

If α is the original significance level, the Bonferroni corrected level is given by

$$\alpha^* = \frac{\alpha}{m} \tag{3.3}$$

Substituting α^* into equation (3.2) generates a probability that is close to (but not equal to unless $m = 1$) α regardless of m.[2] However, the Bonferroni correction ignores the degree of dependency across the tests being performed so that a more general correction, allowing different degrees of dependency among the parameter estimates, can be defined for the parameter estimates associated with covariate j as

$$\alpha_j^* = \alpha / \left[m + D_j (1-m) \right] \tag{3.4}$$

When $D_j = 0$ (no parameter dependency with the model being independently calibrated at each location), the correction reduces to α/m—the original Bonferroni correction. When $D_j = 1$ (complete parameter dependency, and the model is a global one), the significance level will be α (there is no issue of multiple hypothesis testing). When $0 < D_j < 1$ (the usual situation in the calibration of a local model), the corrected significance level is

$$\alpha_j^* = \frac{\alpha}{\left[m + D_j (1-m) \right]} = \frac{\alpha}{\left[m + \frac{m - ENP_j}{m-1}(1-m) \right]} = \alpha / ENP_j \tag{3.5}$$

The dependency in the tests is controlled by the effective number of parameters ENP_j, and this reduces to the solution proposed in da Silva and Fotheringham (2016).

3.2.4 Example 2: Modified Šidák (FWER)

If α is the original significance level, the Šidák corrected level is given by

$$\alpha^* = 1 - (1-\alpha)^{1/m} \tag{3.6}$$

In a similar vein to the Bonferroni modification, to allow for different degrees of dependency among the parameter estimates, define the corrected significance level as

$$\alpha_j^* = 1 - (1-\alpha)^{1/\left(m + D_j (1-m)\right)} \tag{3.7}$$

When $D_j = 0$, the correction is $1-(1-\alpha)^{1/m}$, which is the original Šidák correction. When $D_j = 1$, the significance level is α (there is no issue of multiple hypothesis testing). When $0 < D_j < 1$, the corrected significance level reduces to

$$\alpha_j^* = 1-(1-\alpha)^{1/ENP_j} \tag{3.8}$$

3.2.5 Example 3: Modified Benjamini and Hochberg (FDR)

Unlike the FWER tests that control the proportion of at least one type I error, FDR tests control for the expected proportion of false positives out of all positive discoveries. As a result, FDR increases in statistical power but becomes less conservative at the expense of more false positives. The Benjamini-Hochberg approach is described as follows:

1. Sort the p-values associated with the local parameter estimates from the lowest to highest $p_1 <= p_2 <= ... <= p_m$ and obtain their ranks.
2. Find the largest rank r such that $p_r \leq \alpha \ r / m$.
3. Reject the null hypothesis for all tests at ranks $1...r$.

Note that in FWER each test has the same adjusted p-value, but in FDR each test can have a different adjusted p-value based on its rank among all the tests.

To adjust for the dependency in the tests, the threshold in step 2 is replaced with $p_r \leq \alpha \ r / (m + D(r - m))$. Similar to the dependency-adjusted FWER tests, when D tends to 0 (no dependency), the correction reduces to the classic FDR. When D tends to 1 (global), the correction will apply a global α threshold. When $0 < D_j < 1$, the test will be adjusted based on the dependency in the local parameter estimates.

We now present examples using both simulated and empirical data to demonstrate the differences among the correction methods under the MGWR framework.

3.2.6 Simulated Data Example

Three spatially varying processes $\beta_0, \beta_1, \beta_2$ operating at local, regional, and global scales are simulated onto a 25-by-25 grid having a total of 625 observations. Processes β_0 and β_1 are Gaussian random fields $GRF(0, \Omega)$ with mean of 0 and covariance of Ω, which is denoted as

$$\Omega(h) = exp\left(-0.5 * (d / h)^2\right) \tag{3.9}$$

where d is an m-by-m matrix containing pairwise distances for all locations, and h is a scale parameter indicating the amount of distance-decay in the covariance function. Process β_0 is generated with $h = 10$ and operates at a local scale, process β_1 is simulated with $h = 40$ yielding regional spatial variation, and process β_2 is constant with mean 1 and no spatial variation, representing a global process. The GRF process surfaces are constructed using the *gstools* python package (Müller et al., 2022) and are shown in Figure 3.2.

A simple model is specified in equation (3.10), where the covariates and the error terms follow a normal distribution of *N (0, 1)*.

$$y = \beta_0 + \beta_1 x_1 + \beta_2 x_2 + \varepsilon \tag{3.10}$$

We fit an MGWR model using adaptive bi-square kernel based on the data generating process in equation (3.10) and obtain parameter estimates and uncertainties as well as local t-values. The covariate-specific effective number of parameters is computed and used to adjust the dependency of the tests. A cross comparison of uncorrected t-test (using nominal $\alpha = 0.05$), classic multiple testing adjusted t-tests, and MGWR dependency-controlled t-tests are shown in Figure 3.3. The significant (at the adjusted 0.05 level) local parameters are shown in color while insignificant local parameter estimates are masked in gray.

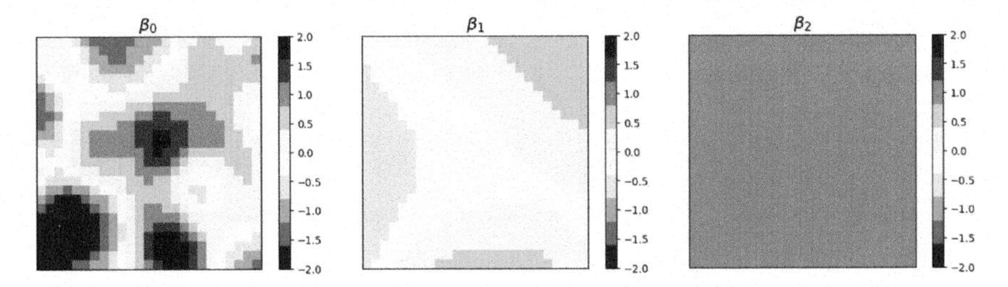

FIGURE 3.2
Three Simulated GRF Parameter Surfaces.

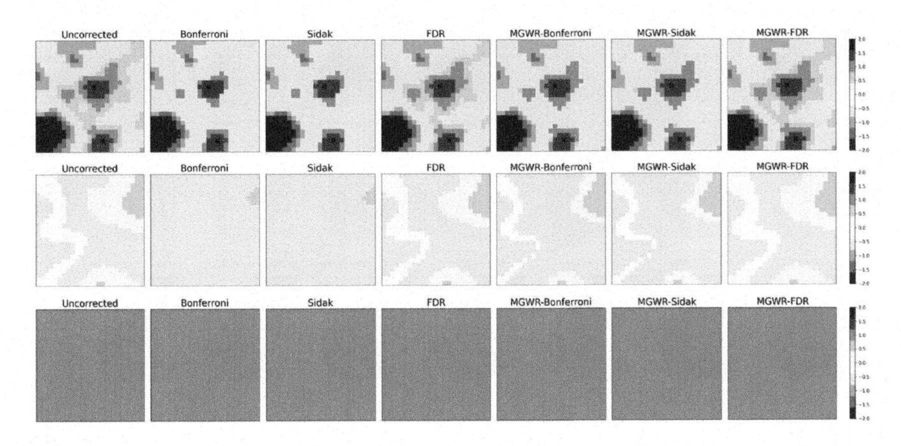

FIGURE 3.3
Illustrations of Various Multiple Testing Correction Methods.

For the two local processes shown in the first and second rows of Figure 3.3, the classic Bonferroni and Šidák corrections show fewer significant local parameter estimates compared to the uncorrected situations, but, without correcting for dependency between the tests, they are overly conservative and mask some interesting results as shown in the MGWR-Bonferroni and MGWR-Šidák tests. Interestingly, both the FDR and MGWR-FDR tests produce very similar results to the uncorrected situation. For the global surface in the last row, all tests reduce to the nominal 0.05 level, and in this example the parameter estimate is statistically significant.

3.2.7 Empirical Data Example

To further compare the performance of these various corrected and uncorrected inferential tests, we employ the Georgia dataset, a test dataset distributed with MGWR 2.2 and for which the data description and download link can be accessed at https://sgsup.asu.edu/sparc/mgwr. An MGWR model is fitted using the county-level education attainment level as the response regressed against three demographic covariates: the percentage of Black residents (PctBlack), the percentage of foreign-born residents (PctFB), and the percentage of rural population (PnctRural) in each county. The local parameters are mapped in Figure 3.4, where the significant ($p < 0.05$) parameters are shown in color, and the insignificant ones are shown in gray.

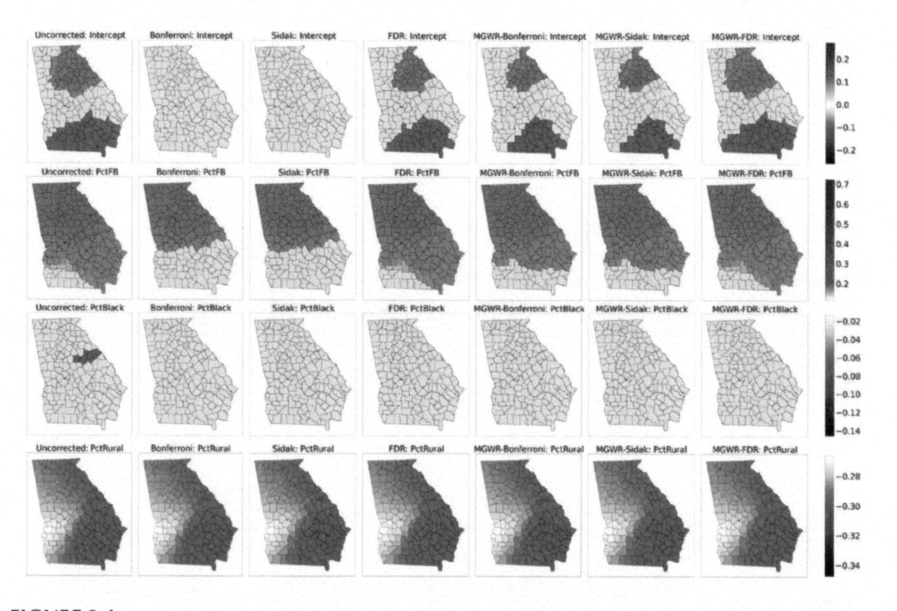

FIGURE 3.4
Multiple Hypothesis Corrections Applied to Parameter Estimates Obtained From the Georgia Example Dataset.

The local intercept maps show clearly the overly conservative nature of the two FWER corrections methods, Bonferroni and Šidák. In both cases, no significant parameter estimates are reported, and type II errors are being committed by not accounting for dependency among the tests. The FDR test, the MGWR-Bonferroni, and the MGWR-Šidák tests are all in agreement and suggest there are two regions where a significant contextual effect is occurring. In the northeast, around Athens where the University of Georgia is located, there is a raised level of the proportion of people with a bachelor's degree, holding the three covariates constant. In the southern counties, bordering Florida, there is a lower-than-expected proportion of people with a degree, accounting for the three covariates. Interestingly, the MGWR-FDR test produces results similar to the uncorrected test showing similar clusters but with more counties in both clusters having significant local intercepts.

For the local parameter estimates associated with the percentage of foreign-born population in each county, the general result is that in the northern counties there is a significant positive association, but this association is not significant in the southern counties. However, where the line is drawn between these two contrasting results depends on the significance test applied. The uncorrected (classical) test and also the FDR and MGWR-FDR tests produce almost identical results with only a few counties in the southwest not having significant local parameter estimates. The Bonferroni and Šidák tests, uncorrected for dependency, are again too conservative, and all the southern counties now have insignificant estimates. The MGWR-Bonferroni and MGWR-Šidák tests produce results in between these two extremes.

For the conditioned association between the percentage of Black residents in each county and education attainment, the results for all the corrected tests (with and without dependency) indicate that there are no locally significant parameter estimates. However, in the uncorrected case, there is a small cluster of significantly negative estimates in the northeast. This is a good illustration of the problem with multiple testing and the lack of correcting for it. Using a 5% probability of committing a type I error, out of 159 counties, there will be on average 8 significant local estimates even when the true relationship is not significant. The adjusted tests all filter out this false signal.

For the conditioned association between the percentage of rural population and education attainment, the relationship is significantly negative for all counties, with the effect being strongest in the southeastern counties and weakest in the West, no matter what test is used.

3.3 Inference About the Overall Spatial Variability of the *Set* of Local Parameter Estimates

Of interest in local modeling is to see whether the modeled spatial variability of a process is significantly different from that expected if the process were stationary. To do so, a null hypothesis is established that all the local parameters are the same and that any variation in their estimates is due solely to sampling variation. To examine rejection or acceptance of this null, a Monte Carlo test is described in Fotheringham et al. (2002) based on GWR. The test shuffles the observations in space (keeping their data intact), re-calibrates GWR on the randomized data while holding the model specification constant, and then computes the variability of the resulting parameter estimates. This process is repeated, and the number of times that the variability of each surface from the randomized data is higher than the variability of the original surface is used to construct pseudo *t*-values for hypothesis testing. A pseudo *t*-value smaller than 0.05 indicates that the observed spatial variability of a parameter estimate surface is significant at the 95% confidence level so that the null hypothesis that the original surface is stationary globally can be rejected in favor of the alternative hypothesis that the original surface is nonstationary. In contrast, a pseudo *t*-value larger than .05 indicates that the null hypothesis cannot be rejected, suggesting that the spatial variation in the original surface of parameter estimates is not significantly different from that expected due to sampling variation, and hence the process generating the local parameter estimates is accepted as being stationary over space. This approach can be easily extended to the MGWR framework using the following steps:

1. Fit an MGWR model with optimal covariate-specific bandwidths and obtain the initial standard deviation of the local parameter estimates for each covariate *j*.
2. Iterate over N ($N >= 1{,}000$) times:
    a. Randomly shuffle the coordinates of the locations for which data are recorded (the values of the covariates and dependent variable remain unchanged).
    b. Re-fit the MGWR model with bandwidth selection for the spatially permutated data using the same settings as in the initial model.
    c. Calculate and store the standard deviation of the local parameter estimates for each covariate.
3. Obtain the empirical *p*-value for each covariate *j* based on the proportion of standard deviations that are above in the list of standard deviations for the permuted data ranked high to low.

To demonstrate the Monte Carlo spatial variability test in MGWR, we perform the test based on the simulated dataset described in Section 3.2.6 using the model described by equation (3.10). The sampling distributions for the spatial variability test for the three sets of local parameters under the null hypothesis are shown in Figure 3.5. The red line denotes the standard deviation of the estimates in the initial model.

For the left-hand (β_0) and the middle (β_1) figures, out of 1,000 iterations, there are 0 instances where the standard deviation of the estimates in the null distribution is higher than the initial model. As a result, the empirical *p*-value is $0/1{,}000 = 0.000 < 0.05$, and the null hypothesis can be rejected so that we infer that both sets of local parameter estimates exhibit significant spatial variability. In the right-hand figure (β_2), out of 1,000

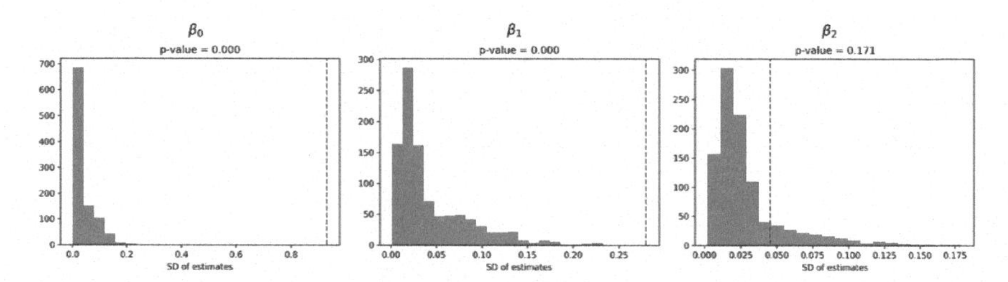

FIGURE 3.5
Sampling Distributions of the Standard Deviation of Local Estimates Under the Null Hypothesis.

iterations, 171 iterations have a higher standard deviation than that for the initial model calibration yielding a p-value of $0.171 > 0.05$, so in this instance, we fail to reject the null hypothesis of no significant spatial variability in the local estimates.

There are alternatives to the Monte Carlo spatial variability test in the case of GWR. For instance, Leung et al. (2000) develop the theoretical sampling distribution of the variance of the parameters under the null hypothesis, and from this p-values can be obtained on the basis of an F-test. However, this has not been extended to MGWR, and the properties of the test statistic have not yet been explored. Another, less-formal, approach is to compare the interquartile range (IQR) of the local estimates to the uncertainty in the equivalent global estimate (Fotheringham et al., 2002) Since the global parameter estimate is the mean of a normal distribution, 68% of the realizations of that estimate will lie within one standard deviation of that value. Consequently, if the IQR of the local estimates, which, by definition, contains 50% of the local estimates, is greater than one standard deviation of the global mean, this supports the possibility that the relationship may not be stationary.

Another approach resorts to model selection to find out whether fixing certain parameters as stationary will produce a better model than allowing all the relationships to be spatially varying. This approach, which was originally used in semi-parametric GWR (Fotheringham et al., 2002), has been superseded by MGWR. A large (relative to the number of observations) covariate-specific optimal bandwidth in MGWR indicates that the conditioned association represented by that bandwidth is stationary over space. We explore this approach further later.

3.4 Inference About the Bandwidth

3.4.1 Bias-Variance Trade-Off

Bandwidth is a key parameter in the data-borrowing methodology behind MGWR. As described in Chapter 2, bandwidth controls the bias-variance trade-off in modeling each spatially varying process. Consider a *true* parameter surface as shown in Figure 3.6. There exists an *optimal* bandwidth that can be used to best approximate

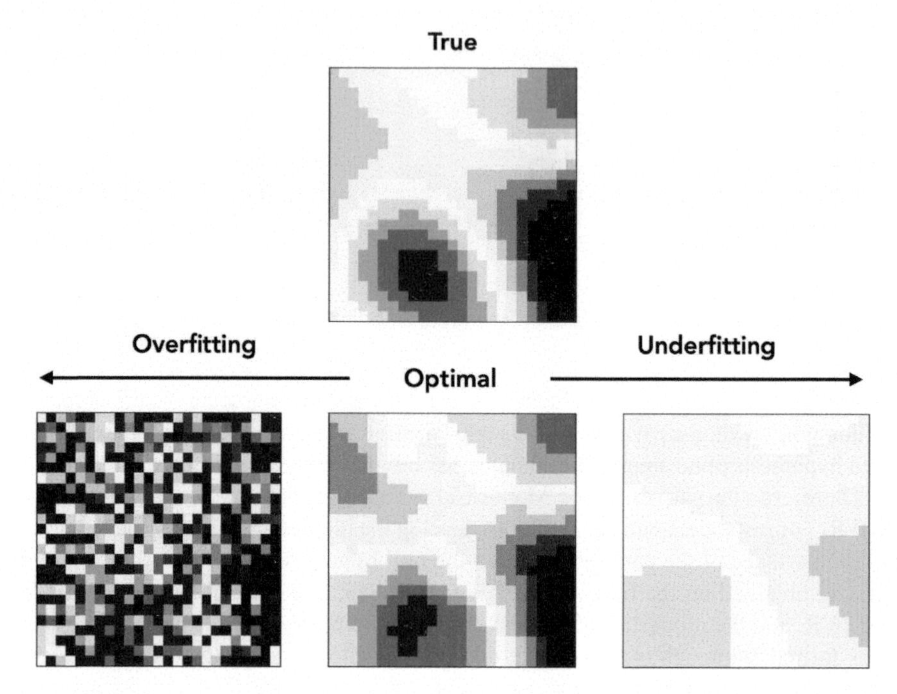

FIGURE 3.6
An Illustration of How an Optimal Bandwidth Balances Overfitting and Underfitting.

the spatial pattern in the true surface. When the bandwidth is small, fewer data points are used in each local regression, and the local parameter estimates will be more sensitive to sampling noise in the data. When the bandwidth is large, more data points are included in each local regression so that the local parameter estimates will be more precise but will be more prone to bias because data are used in the local regression that are more likely to have been produced by different processes to those being estimated. Accordingly, using a smaller-than-optimal bandwidth will likely overfit the data while a larger-than-optimal bandwidth will underfit the data as shown in Figure 3.6. Finding the set of optimal, covariate-specific bandwidths is therefore critical in MGWR to both accurately and precisely model the underlying spatial processes.

Finding the optimal bandwidth is essentially a model selection problem. The optimal bandwidth is often chosen based on cross-validation statistics or information criteria. Information theory-based approaches such as *AIC* (see more details in Chapter 2) offer an intuitive interpretation as a trade-off between model fit and model complexity as well as allowing evaluation of model selection uncertainty. It is suggested that when the sample size is small, *AICc* should be used instead of *AIC*, and *AICc* is equivalent to *AIC* when the sample size is large (Hurvich & Tsai, 1989). Figure 3.7 demonstrates the use of *AICc* to determine an optimal bandwidth. The red vertical line marks the optimal bandwidth from the bandwidth selection routine in MGWR with the smallest *AICc* value that yields a best model.

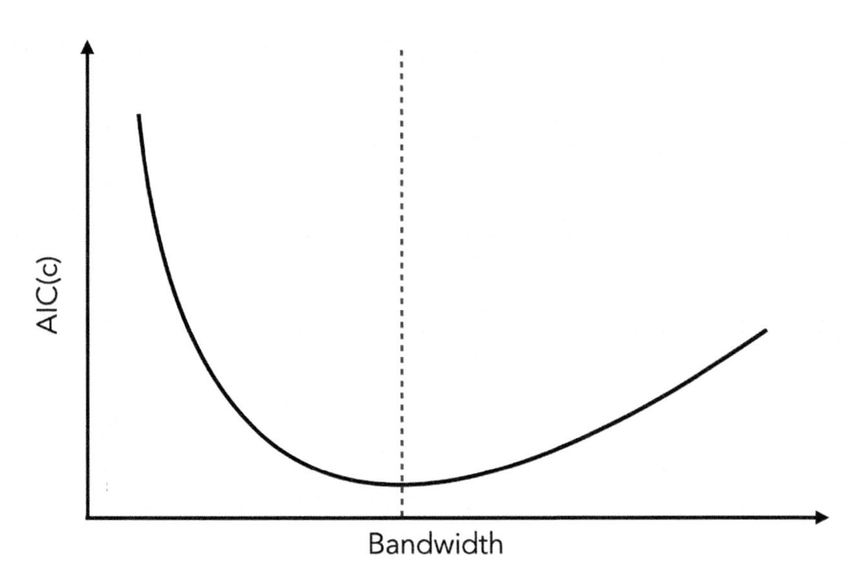

FIGURE 3.7
The Relationship Between Bandwidth and *AICc*.

It is worth noting that *AICc* also asymptotically selects the model-minimizing mean-square error of an estimate (Yang, 2005; Vrieze, 2012), which can be conveniently decomposed into the sum of the variance and the square of the bias of the estimate as shown in equation (3.11).

$$\begin{aligned} MSE\left(\hat{\beta}_i\right) &= E\left[\left(\hat{\beta}_i - \beta_i\right)^2\right] \\ &= Var\left(\hat{\beta}_i - \beta_i\right) + \left(E\left[\hat{\beta}_i - \beta_i\right]\right)^2 \\ &= Var\left(\hat{\beta}_i\right) + Bias^2\left(\hat{\beta}_i\right) \end{aligned} \tag{3.11}$$

Figure 3.7, which is equivalent to Figure 1.5, illustrates the relationship between squared bias and variance in the local parameter estimates and bandwidth. As the bandwidth increases, bias in the local parameter estimates increases while the variance of the estimates decreases. While the variance flattens out quite rapidly, the bias increases continuously as the bandwidth increases. There exists a point where the increase in the squared bias is offset exactly by a decrease in the variance, and this is the optimal bandwidth (shown by the dashed line), which can be determined more easily by finding the minimum *AICc* value (Yu et al., 2020a, 2020b). Note that the optimal bandwidth is selected based on only one realization of the data, and it is subject to sampling variation and model selection uncertainty, so it is more likely that an empirical optimal bandwidth will fall within a wider confidence region, which is illustrated in Figure 3.8 by the gray area.

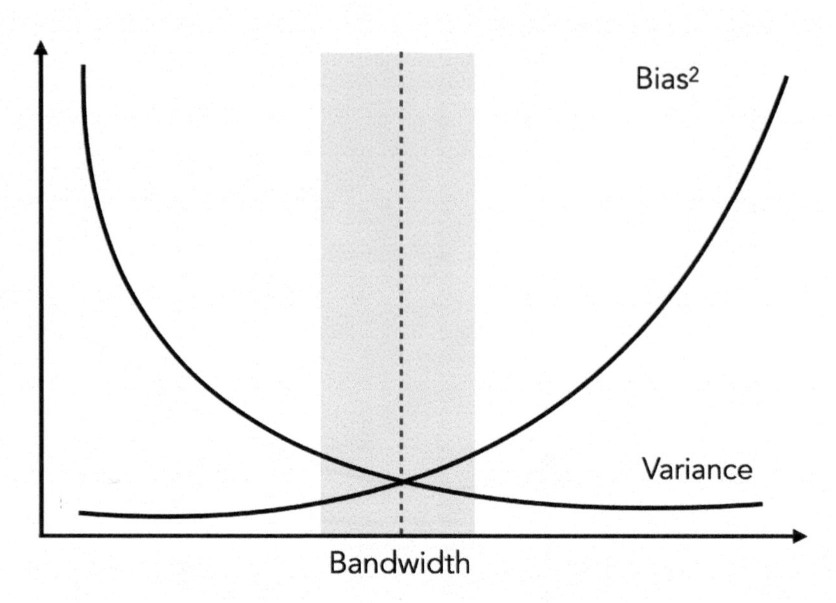

FIGURE 3.8
An Illustration of How an Optimal Bandwidth Balances Bias and Variance.

3.4.2 Measuring Bandwidth Uncertainty

When multiple realizations of data are available, bandwidths will vary based on the sampling variation of the data used for the model calibration. It is important to know how stable or sensitive the optimal bandwidth is to sampling variation in order to make statements about the nature of the process variability described by the bandwidth and also to compare two or more bandwidths. The analytical distribution of the optimal bandwidth is unknown, but there are computational approaches to approximate this distribution. One is to use the bootstrap method detailed in Efron (1987). This involves the following steps:

1. Calibrate an MGWR model with optimal covariate-specific bandwidths and obtain the model predictions and residuals.
2. Iterate over N ($N> = 1{,}000$) times:
    a. Randomly resample residuals with replacement and add to the model prediction as the new dependent variable.
    b. Fit an MGWR model with this new dependent variable and retain the optimal bandwidths.
3. Obtain a confidence interval for each bandwidth using the percentile method based on the empirical bootstrap distribution of bandwidths.

Following this approach, we are able to obtain bootstrap bandwidth distributions. As an example, Figure 3.8 shows the bootstrap bandwidth distributions for the covariate-specific optimal bandwidths using the simulated dataset described in Section 3.2. The distribution for the local bandwidth bw_0 is quite narrow, while the distribution for bw_1 is wider. The bandwidth bw_2, which indicates a global relationship, is mostly at the maximum possible value of 625 nearest neighbors (there are 625 data points), but it can be as low as 100. This indicates that the optimal bandwidth can be much smaller than the number of locations at which a local regression is performed but still indicates a global relationship. The bandwidth distribution also suggests that the bandwidths are not deterministic values but stochastic and are subject to sampling variations. Ninety-five percent bandwidth confidence intervals can be obtained by computing the bandwidths at the 2.5% and 97.5% percentiles of the distributions. The bandwidth confidence intervals for these example datasets are [30, 50], [60, 286], and [124, 625] for the three covariates, respectively. These are shown by the gray areas in Figure 3.9.

Another computationally more efficient approach to measuring bandwidth uncertainty is based on model selection uncertainty and multi-model inference, which is well discussed in Burnham and Anderson (1998). For *AIC*-based model selection, Akaike weights can be used to measure model selection uncertainties. In the context of MGWR, candidate bandwidths are evaluated based on *AICc* (or an equivalent measure), and the bandwidth with the minimum *AICc* is selected as the optimal one. It is a natural extension therefore to use the Akaike weights to measure the relative likelihood of a bandwidth being selected as optimal in order to quantify the bandwidth selection uncertainty. This approach has been shown to be useful elsewhere (Burnham & Anderson, 1998; Wagenmakers & Farrell, 2004; Symonds & Moussalli, 2011; Burnham et al., 2011; Posada & Buckley, 2004; Johnson & Omland, 2004; Koh & Wilcove, 2008; Pinsky et al., 2013).

Li et al. (2020) extend the use of Akaike weights to measure bandwidth selection uncertainties within the MGWR framework. To calculate Akaike weights, a set of competing models is defined, and in the (M)GWR framework, these competing models are defined by R candidate bandwidths $\{bw_1, bw_2 \ldots bw_R\}$. The minimum AIC using the R candidate bandwidths is denoted as AIC_{min}. For bandwidth k within the candidate set R, an AIC difference can be computed as $\Delta_k = AIC_k - AIC_{min}$. Then, the Akaike weight of a candidate bandwidth $k \in \{1\ldots, R\}$ can be obtained by

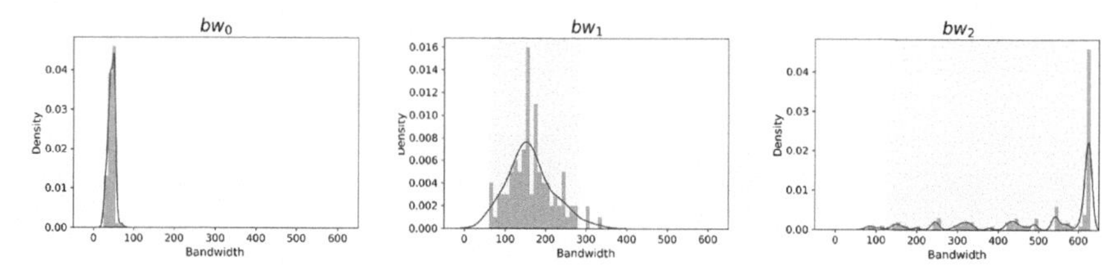

FIGURE 3.9
Bootstrapped Bandwidth Distributions.

$$w_k = \frac{exp\left(-\frac{1}{2}\Delta_k\right)}{\sum_{r=1}^{R} exp\left(-\frac{1}{2}\Delta_r\right)} \tag{3.12}$$

Given maximum likelihood estimators based on the same data, the numerator $exp\left(-\frac{1}{2}\Delta_k\right)$ represents the likelihood of the bandwidth *k*, which also quantifies the relative strength of evidence for each bandwidth (Akaike, 1981; Burnham & Anderson, 1998). The Akaike weights are normalized using the denominator so that the range of values is 0 to 1, with the sum being 1 ($\sum w_k = 1$). The likelihood that a particular bandwidth is the optimal one is represented by the resulting Akaike weight. A bandwidth with an Akaike weight of 0.6 has a 60% chance of being chosen as the best bandwidth among the defined candidate bandwidths.

According to Symonds and Moussalli (2011) and Burnham and Anderson (1998), a 95% confidence interval (CI) of bandwidths can be created by ranking the Akaike weights in descending order, and by adding bandwidths to the CI until, the cumulative Akaike weight reaches 0.95. When a crude searching routine is used, the cumulative Akaike weight might not be exactly 0.95, and in such cases the inclusion to the bandwidth CI should stop when the cumulative Akaike weight is just above 0.95. Table 3.1 shows how this procedure operates for modeling the local spatial process β_1 in Figure 3.2. An interval bandwidth search with a step of 20 is

TABLE 3.1

An Example of Obtaining the 95% Confidence Set of Bandwidths Using Akaike Weights.

Bandwidth	*AICc*	Akaike weight	Cum. Akaike weight
140	**1742.99**	**0.17**	**0.17**
160	1743.12	0.16	0.33
120	1743.36	0.14	0.48
180	1743.65	0.12	0.60
200	1744.15	0.10	0.70
100	1744.32	0.09	0.79
220	1745.22	0.06	0.84
240	1745.76	0.04	0.89
260	1746.31	0.03	0.92
80	1747.27	0.02	0.94
280	1747.28	0.02	0.96
300	1748.07	0.01	0.97
320	1748.88	0.01	0.98
340	1749.85	0.01	0.99
360	1750.74	0.00	0.99
...	...	...	...

used here. The optimal bandwidth is selected as 140, and it has an Akaike weight of 0.17. Bandwidths of 160, 120, and 180 have descending Akaike weights of 0.16, 0.14, and 0.12, respectively. Inclusion into the 95% CI stops after the addition of the bandwidth 280 because at this point the cumulative Akaike weight marginally exceeds 0.95 (0.96). Consequently, we can state that the 95% (more accurately, 96%) CI of the bandwidth is [80, 280]. Higher precision can be gained by using a smaller step size in the bandwidth search. It should be noted that the bandwidth CI is not expected to be symmetrical or continuous. It is also possible that with a certain arrangement of the spatial data or a pattern of the underlying process that the resulting bandwidth distribution could be multimodal. The Akaike weights-based CI is similar to the bootstrap CI of [60, 286], but it has the advantage of not requiring repetitive computation. The bandwidths' CIs can be readily obtained with no additional computation involved. Following the same approach, Akaike weights-based CI for the other two bandwidths, bw_0 and bw_2, are [40, 60] and [260, 625], respectively.

The bandwidth confidence intervals can be used for inference in two ways. The first is to infer the differences in the operating scale of the process as measured by the bandwidth. If the bandwidths of two processes have distinctly non-overlapping confidence intervals (e.g., [40, 60] for β_0 versus [260, 625] for β_2), we can state that they operate at two statistically different spatial scales based on the data at hand. It is worth noting that the opposite may not always be true if two intervals are overlapping; there is still a possibility that the bandwidths are statistically different depending on the variability of the two distributions being compared (see more discussion regarding examining overlapping CIs in Schenker & Gentleman, 2001). The second is to infer the spatial stationarity of the parameter estimate surface by comparing the confidence interval with the upper limit of the bandwidth. In this example, the upper limit of the bandwidth is the total number of data points, which is 625. Based on the fact that the bandwidth confidence interval of β_2 includes 625, we can state that the parameter estimates for β_2 are not statistically different from being stationary, while the other two estimated processes exhibit significant spatial variability.

3.5 Summary

In this chapter, three main inferential issues in MGWR are described: (i) inference about the *individual* local parameter estimates; (ii) inference about the overall spatial variability of the *set* of local estimates; and (iii) inference about the covariate-specific optimal *bandwidths*. In local model calibration, classic inference is complicated by multiple hypothesis testing and by the dependency in the parameter estimates so that corrections have to be applied. Standard correction procedures for multiple hypothesis testing have themselves to be corrected to account for the degree of dependency between the tests. We show how this is easily done within the MGWR framework. However, there is more research needed into developing analytical distributions within the MGWR framework for significance testing, particularly for making inferences about surfaces of parameter estimates. The current reliance on Monte Carlo distributions comes with a heavy computational burden.

Notes

1. The rate of increase of D could also be controlled by a log function reflecting the increased sensitivity of the degree of dependency in the local parameter estimates to variations in the optimal bandwidth at small values of the bandwidth. Such a measure of dependency can be termed D^* and is calculated as

$$D_j^* = \frac{log(m) - log(ENP_j)}{log(m)}$$

   This alternative measure of dependency is also bounded between 0 and 1, but it has a more linear rate of change with changes in bandwidth. In MGWR 2.2, both D and D^* are reported.
2. The values decline as m increases but far less quickly than if the test is uncorrected.

References

Akaike, H. (1981). Likelihood of a model and information criteria. *Journal of Econometrics, 16*, 3–14.

Benjamini, Y., & Hochberg, Y. (1995). Controlling the false discovery rate: A practical and powerful approach to multiple testing. *Journal of the Royal Statistical Society: Series B (Methodological), 57*(1), 289–300.

Bonferroni, C. (1936). Teoria statistica delle classi e calcolo delle probabilita. *Publicazioni del R Istituto Superiore di Scienze Economiche e Commericiali di Firenze, 8*, 3–62.

Burnham, K. P., & Anderson, D. R. (1998). *Model selection and inference: A practical information-theoretic approach*. New York: Springer.

Burnham, K. P., Anderson, D. R., & Huyvaert, K. P. (2011). AIC model selection and multimodel inference in behavioral ecology: Some background, observations, and comparisons. *Behavioral Ecology and Sociobiology, 65*(1), 23–35.

da Silva, A. R., & Fotheringham, A. S. (2016). The multiple testing issue in geographically weighted regression. *Geographical Analysis, 48*(3), 233–247.

Efron, B. (1987). Better bootstrap confidence intervals. *Journal of the American Statistical Association, 82*(397), 171–185.

Fotheringham, A. S. (2023). A comment on 'A route map for successful applications of geographically-weighted regression': The alternative expressway to defensible regression-based local modelling. *Geographical Analysis, 55*(1), 191–197.

Fotheringham, A. S., Brunsdon, C., & Charlton, M. E. (2002). *Geographically weighted regression: The analysis of spatially varying relationships*. London: Wiley.
Hurvich, C. M., & Tsai, C. L. (1989). Regression and time series model selection in small samples. *Biometrika*, *76*(2), 297–307.
Johnson, J. B., & Omland, K. S. (2004). Model selection in ecology and evolution. *Trends in Ecology & Evolution*, *19*(2), 101–108.
Koh, L. P., & Wilcove, D. S. (2008). Is oil palm agriculture really destroying tropical biodiversity? *Conservation Letters*, *1*(2), 60–64.
Leung, Y., Mei, C. L., & Zhang, W. X. (2000). Statistical tests for spatial nonstationarity based on the geographically weighted regression model. *Environment and Planning A*, *32*(1), 9–32.
Li, Z., Fotheringham, A. S., Oshan, T. M., & Wolf, L. J. (2020). Measuring bandwidth uncertainty in multiscale geographically weighted regression using Akaike weights. *Annals of the American Association of Geographers*, *110*(5), 1500–1520.
Müller, S., Schüler, L., Zech, A., & Heße, F. (2022). GSTools v1.3: A toolbox for geostatistical modelling in Python. *Geoscientific Model Development*, *15*(7), 3161–3182.
Pinsky, M. L., Worm, B., Fogarty, M. J., Sarmiento, J. L., & Levin, S. A. (2013). Marine taxa track local climate velocities. *Science*, *341*(6151), 1239–1242.
Posada, D., & Buckley, T. R. (2004). Model selection and model averaging in phylogenetics: Advantages of Akaike information criterion and Bayesian approaches over likelihood ratio tests. *Systematic Biology*, *53*(5), 793–808.
Schenker, N., & Gentleman, J. F. (2001). On judging the significance of differences by examining the overlap between confidence intervals. *The American Statistician*, *55*(3), 182–186.
Šidák, Z. (1967). Rectangular confidence regions for the means of multivariate normal distributions. *Journal of the American Statistical Association*, *62*(318), 626–633.
Symonds, M. R., & Moussalli, A. (2011). A brief guide to model selection, multimodel inference and model averaging in behavioural ecology using Akaike's information criterion. *Behavioral Ecology and Sociobiology*, *65*(1), 13–21.
Vrieze, S. I. (2012). Model selection and psychological theory: A discussion of the differences between the Akaike information criterion (AIC) and the Bayesian information criterion (BIC). *Psychological Methods*, *17*(2), 228.
Wagenmakers, E. J., & Farrell, S. (2004). AIC model selection using Akaike weights. *Psychonomic Bulletin & Review*, *11*(1), 192–196.
Yang, Y. (2005). Can the strengths of AIC and BIC be shared? A conflict between model identification and regression estimation. *Biometrika*, *92*(4), 937–950.
Yu, H., Fotheringham, A. S., Li, Z., Oshan, T. M., Kang, W., & Wolf, L. J. (2020b). Inference in multiscale geographically weighted regression. *Geographical Analysis*, *52*(1), 87–106.
Yu, H., Fotheringham, A. S., Li, Z., Oshan, T. M., & Wolf, L. J. (2020a). On the measurement of bias in geographically weighted regression models. *Spatial Statistics*, *38*, 100453.

4

Spatial Scale and Local Modeling

4.1 Setting the Scene

There is a strong connection between the concepts of spatial scale, spatial heterogeneity, local models, and context, as shown in Figure 4.1. Namely, in order to measure and understand the context of a spatial process, there must be some spatial heterogeneity associated with the process, which requires a local model to account for such heterogeneity, and central to local models is the spatial scale(s) at which variations occur and are measured. Without spatial heterogeneity among spatial processes there would be no need to study different contexts; without local models it would not be possible to recognize and measure spatially heterogeneous processes; without the notion of spatial scale there would be no intuition about the nature of spatially heterogeneous processes and the relative scope of different contexts. Chapter 1 introduced the ideas of spatial heterogeneity, context, and local models. This chapter focuses on the fundamental role of spatial scale within context-driven local models.

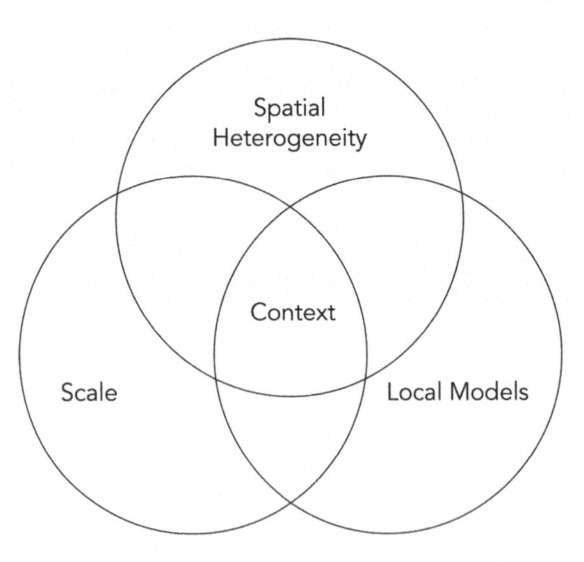

FIGURE 4.1
Interconnectivity Between Scale, Spatial Heterogeneity, Local Models, and Context.

DOI: 10.1201/9781003435464-4

Scale is a universal notion across the sciences for arranging the world around us into concepts, entities, and interactions that exist along a spectrum. It is therefore the lens that determines when things can be perceived and allows entities to be grouped together or differentiated from one another. Different scales can be used to organize different dimensions of the world, such as the temporal scale for chronological organization or measurement scales for organizing temperature or weight observations. In this chapter, the focus is on the spatial scale, which has been widely studied and is used broadly to organize geographic concepts, entities, and interactions, or those that can be perceived and referenced in relation to the Earth's surface (Meentemeyer, 1989; Lam & Quattrochi, 1992; Marceau, 1999; Gibson et al., 2000; Atkinson & Tate, 2000; Goodchild, 2001; Dungan et al., 2002; Wu & Li, 2006; Manson, 2008; Ruddell & Wentz, 2009; Wu & Li, 2009; Dabiri & Blaschke, 2019). Given the broad applicability of spatial scale, this chapter briefly introduces how the concept of scale has developed in the field of spatial analysis and then expands specifically on the role of spatial scale in context-driven modeling. After outlining several types of spatial scale and how leveraging multiple scales drives spatial analytical modeling, the concept of an *indicator of scale* is introduced and discussed for the example of MGWR, and some additional scale issues are highlighted from the perspective of local modeling.

4.2 Scale in Spatial Analysis

4.2.1 Types of Spatial Scale

Though spatial scale manifests in many ways, there are four predominant variations that are most frequently encountered in spatial analysis (Dabiri & Blaschke, 2019; Oshan et al., 2022). The first is cartographic scale, which relates to the amount of detail represented by a map, either digital or analog (Goodchild, 2001). Here, scales are often measured by a representative fraction that is a ratio between detail in the real world and detail on a map. A small scale map has a representative fraction with a small denominator and contains a high level of detail or information whereas a large scale map has a representative fraction with a large denominator and therefore contains a lower level of detail or information. In contrast, the second and third types of scale, referred to as the geographic scale (or extent) and observation scale (or grain), adopt the inverse classification with large scales relating to more detail, information, and areal coverage. The former is used to describe the boundaries of a study area or a spatial sample of interest while the latter is used to describe the units of analysis (Goodchild, 2001; Dungan et al., 2002). Both of these types of scale characterize facets of the data used in spatial analysis, and though the specification of these two types of scales is sometimes based on theory, it is more often based on practical decisions or convenience. The primary scale of interest in context-driven modeling and the fourth and final type discussed here—that of process (or phenomena) scale—is different in that it characterizes relationships, events, and entities in the real world rather than the data we collect. Thus, process scale is inherently a theoretical construct and requires the use of data-centric scales or additional spatial concepts (such as distance, lag, dependence, cluster, hierarchy) in order to be described. Examples of process scales include the range at which we find samples that are similar to each other, the level of the atmosphere where we find the formation of certain weather features, or the average size of a particular tree species within a larger stand of trees that is part of a forest.

There exist several additional scale terminologies, many of which are synonymous with the concepts described here. For instance, 'resolution' is another term often used to describe the observation scale of a study area, especially when using remotely sensed data collected across uniform tessellations (Marceau & Hay, 1999). And other types of scale appear in the literature although much less frequently, such as policy scale, which describes the scale that decisions can be made, or analysis scale, which describes a common scale used to harmonize model input data sources (Dabiri & Blaschke, 2019). An issue that arises in practice is that sometimes the same terminologies are used to refer to different conceptualizations or that different terms are used to refer to the same conceptualization (Silbernagel, 1997; Csillag et al., 2000; Jenerette & Wu, 2000; Oshan et al., 2022). It is therefore important to avoid potential ambiguities, and an effort is made here to introduce only a core subset of scale concepts and terms rather than provide an exhaustive overview. This specificity becomes even more important when considering multiple simultaneous scales and the array of spatial analysis techniques that leverage them.

4.2.2 From One Scale to Multiple Scales

Defining and using at least one scale is a necessary requirement for geographic research and most spatial analytical techniques, but the ability to consider or integrate multiple scales simultaneously unlocks powerful analytical capabilities. Oshan et al. (2022) discuss an array of actions and techniques that can enhance research and technology through the use of more than one scale. For example, digital cartography and geographic information systems have focused on developing multiscale data structures and automated processing algorithms so that analysts can efficiently store, access, and render geographic data across many scales. Methods have also been developed for transforming data at one scale to another (i.e., upscaling and downscaling), detecting or creating clusters across scales, extracting features from remotely sensed data, and representing objects and environments at different scales in agent-based modeling. More importantly, the selection and integration of information at different scales are central to context-driven modeling. The most naive method for exploring scale in spatial models consists of obtaining results separately for data at different scales and comparing them. Typically, as shown in Figure 4.2, either the geographic scale or observation scale is varied while the other is held constant. The result is that it becomes possible to learn about the role of scale in generating patterns and processes and identifying the scale at which they become

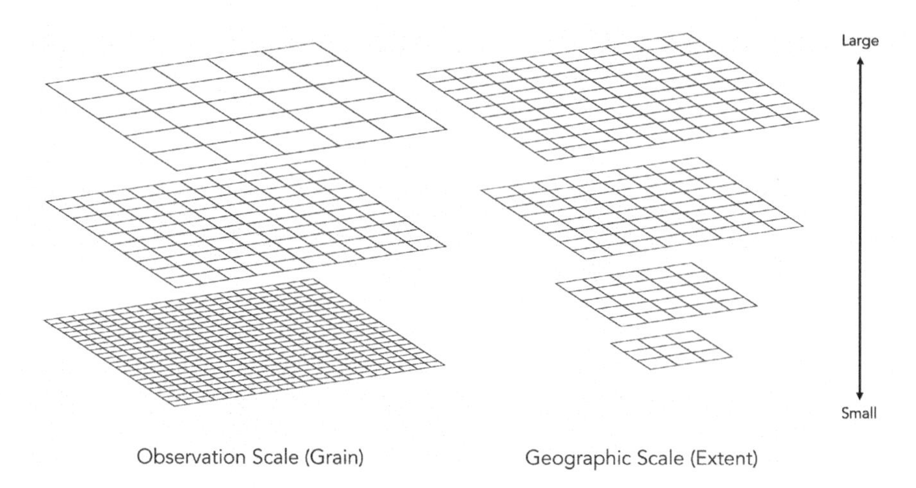

FIGURE 4.2
Multiple Observation Scales (Left) and Multiple Geographic Scales (Right).

most salient. More sophisticated methods for comparing or combining scales, such as MGWR, allow for inference to be conducted and uncertainties to be quantified pertaining to scale parameters, giving rise to the notion of indicators of scale that are at the heart of context-driven modeling.

4.3 Indicators of Scale

4.3.1 Two Paradigms for Investigating Process Scale

It is relatively straightforward to describe the geographic scale used to delineate a study area or the observation scale used to collect samples. In contrast, there is no natural unit to describe the scale(s) of a process so that geographic scale and observation scale are typically used as proxies to describe the range over which a process is relatively constant. An issue that arises is that we do not typically know the appropriate geographic scale or observation scale to describe a process, especially if the process spans multiple scales. Two main paradigms have been employed to uncover information about process scale. The first is hinted at earlier and involves creating a profile to compare a quantity of interest calculated separately at each scale within a range of scales. It then becomes possible to compare the profile to a theoretical distribution as in the case of a variogram (to identify scales where variation is relatively high or low), compute a summary metric such as the entropy of the profile (to characterize complexity across of scales), or select a minimum or maximum of the profile as can be done for spatial autocorrelation statistics (to identify an optimal scale to summarize a clustering pattern).

The second paradigm involves decomposing the variation of a spatial variable as a function of different scales. For instance, multilevel models decompose a variable as a function of itself or other variables at different observation scales, and MGWR decomposes a variable as a function of other variables at different geographic scales. These methods allow the contribution of different scales to the overall variation in a spatial pattern to be assessed and ranked. Both the profiling paradigm and the decomposition paradigm hold either the geographic scale or the observation scale constant while varying the other in order to learn about the scale of spatial processes, whether through informal comparative mechanisms or through more formal inferential means (Oshan et al., 2022). As a result, the methods that fall under these two paradigms may be considered to produce *indicators of scale* or quantities of interest that measure aspects of process scale and allow us to learn about spatial context. MGWR incorporates features of both paradigms in order to produce a sophisticated indicator of scale known as *the bandwidth* for characterizing spatial processes.

4.3.2 Bandwidth, Scale, and the Spatial Weights Matrix

One way of describing process scale is the extent or range of dependence between the relationships or associations generating the data we observe. This is typically captured in a variety of spatial analysis methods through the use of spatial weights matrices (SWMs), including global and local forms of both univariate statistics that measure spatial autocorrelation and spatial regression models that measure conditional associations across a set of covariates. Creating an SWM requires the selection of a fixed observation scale (i.e., spatial support), as well as additional parameters that modulate the neighborhood around each data point for defining potentially related observations, which is equivalent to varying the effective geographic scale for each neighborhood. Thus, the SWM is the primary formal mechanism for representing the spatial dependence theorized by Tobler's first law of geography (Tobler, 1970; Miller, 2004), and the selection of these additional parameters is akin to specifying the scale of the process(es) being measured (Rogerson, 2021). In the case of GWR and MGWR, SWMs are most often defined using a distance threshold or the number of nearest neighbors beyond which observations are considered unrelated, taking values between zero and one in accordance with the notion that more distant relationships should be weaker (i.e., distance-decay). In order to optimize the analysis of spatial processes, the choice of bandwidth is treated as a hyperparameter that can be tuned in order to maximize the magnitude of a statistic and the relationship it represents. This strategy has been demonstrated for both the local Moran's I and the Getis-Ord G_i spatial autocorrelation statistics (Getis & Aldstad, 2004; Aldstad & Getis, 2006; Rogerson & Kedron, 2012; Rogerson, 2015). It is also at the core of the MGWR framework where the SWM is selected by optimizing a bandwidth parameter based on an information criterion that balances a trade-off between bias and variance in the model's local parameter estimates (Fotheringham et al., 2002; Yu et al., 2020), imparting additional meaning on the bandwidth.

4.4 The Bandwidth in MGWR as an Indicator of Scale

4.4.1 The Meaning of Bandwidth in MGWR

In the MGWR framework, the bandwidth captures the neighborhood around a local regression point from which data are weighted and borrowed to enable the calibration of a model at each regression point. Both a failure to borrow data from a large enough neighborhood and including data points beyond the bandwidth result in local parameter estimates that are suboptimal. In addition, the number of nearest neighbors that data are borrowed from can vary across the covariates in the model, providing information on how each separate conditional relationship varies over space. Smaller bandwidths mean that parameters for specific locations can be optimally estimated by using information obtained from very close to those locations and indicate processes that vary relatively rapidly over space; larger bandwidths mean that parameter estimation at each location requires information obtained from a much wider range of locations and indicate processes that vary relatively slowly over space. The bandwidth parameter is therefore an important output in MGWR because it indicates how local or global each process being modeled is (Fotheringham et al., 2022).

There are two primary ways to understand how the bandwidth acquires the above interpretation. The first is through the lens of a bias-variance trade-off that has been explored in MGWR (Yu et al., 2020). If the bandwidth is large, data from more sites are included in the local calibration whereas if the bandwidth is small, data from only a relatively few nearby sites are used in the calibration. As more distant locations are added to the local regression, bias in the local parameter estimates increases because the relationships these additional data represent are more likely to differ from those at the regression point. At the same time, adding more data to the local regression will decrease the standard errors of the local parameters as the local parameters will be estimated on the basis of larger samples. The bandwidths produced by an MGWR calibration are therefore optimal in the sense that each denotes the number of locations (or distance from the regression point) beyond which adding data will increase the bias in the parameter estimates more than it will decrease the uncertainty about the estimates.

The second way to understand the bandwidth in MGWR is through the concept of (mis)information. Local regressions can be calibrated through MGWR by borrowing information from locations that are weighted according to their proximity to the point where the local regression is being carried out. Data at each location provide some information but also some misinformation on the processes operating at the local regression point. The locations up to the bandwidth provide data that contain more information than misinformation on the processes at the local regression point and are therefore included in the calibration. Beyond the bandwidth, data contain more misinformation than information on the processes at the local regression point and are therefore excluded from the calibration. MGWR is able to accommodate an SWM parameterized with a separate bandwidth for each process in the model so that the above interpretations can be explored in a relationship-wise manner. In many applications, some relationships will have relatively small associated bandwidths, and other relationships will have relatively large bandwidths. In the case of (mis) information, a small bandwidth indicates that as distance from the local regression point increases, the amount of misinformation in the data quickly exceeds the amount of information. In contrast, a large bandwidth indicates that as distance from the regression point increases the levels of misinformation tend to increase very slowly so that larger numbers of data points are included in the local regressions to reduce parameter estimate uncertainty without much penalty on the overall amount of information contained in the samples. In theory, relationships that have no spatial variation will have infinitely large bandwidths because adding data will reduce the standard errors of the parameter estimates without introducing any misinformation. This same logic applies when interpreting the optimal bandwidth as bias-variance trade-off, and consequently both explanations support the interpretation of the optimized bandwidths that result from MGWR calibration as a summary of the spatial scale over which the relationships vary.

4.4.2 The Determinants of Bandwidth in MGWR

Although the optimized, covariate-specific bandwidths resulting from the calibration of an MGWR model indicate the spatial scale of the processes being examined, the exact determinants of the bandwidth are complex. Fotheringham et al. (2022) examined the sensitivity of bandwidth estimates to three attributes related to the characteristics of processes: (i) local parameter variance; (ii) local parameter spatial dependency; and (iii) local parameter strength. As a process becomes increasingly global in nature (i.e., spatially homogeneous), the local parameter estimates reflecting this process will become increasingly uniform, and the resulting bandwidth will become larger. However, for any given degree of parameter variance, the bandwidth will also be affected by the degree of parameter spatial dependence: a high degree of spatial dependence will generally lead to smaller bandwidths, *ceteris paribus*. Local parameter strength represents the signal captured by the model in comparison to the noise captured by the model error term. Processes that are very weak (with large model error) and where changes in a covariate magnitude have little impact on the magnitude of the dependent variable may, *ceteris paribus*, exhibit different bandwidths from those which are strong.

A series of simulation experiments was used to formally demonstrate the relationship between the optimal bandwidth in MGWR and local parameter variance, spatial dependency, and strength. Chief among these was the generation of synthetic datasets based on local parameter surfaces created using a Gaussian random field with a range of parameter values to control the spatial structure of the surfaces. Assuming a normally distributed covariate and error term and a fixed level of error variance, a dependent variable was calculated, and an optimal bandwidth was estimated for each dataset (in this experiment the model being calibrated had a single covariate so GWR was appropriate). The resulting bandwidth estimates were then organized based on the local parameter standard deviation as a proxy for variance and Moran's I spatial autocorrelation coefficient as a proxy for spatial dependency (Fotheringham et al., 2022).

The results of the experiment demonstrated two dominant trends. First, when parameter spatial dependency (as measured by Moran's I) is held constant, the estimated bandwidth tends to decrease as the variation in the local parameters increases. A generalized version of this trend is illustrated in Figure 4.3. Second, although less sensitive, when the variation in the local parameters is held constant, the estimated bandwidth tends to first decrease as spatial dependency increases up until a moderate-to-high level whereby the bandwidth then starts to increase in the presence of extreme spatial dependency. A generalized version of this trend is illustrated in Figure 4.4, where the smallest bandwidths are associated with moderate-to-high

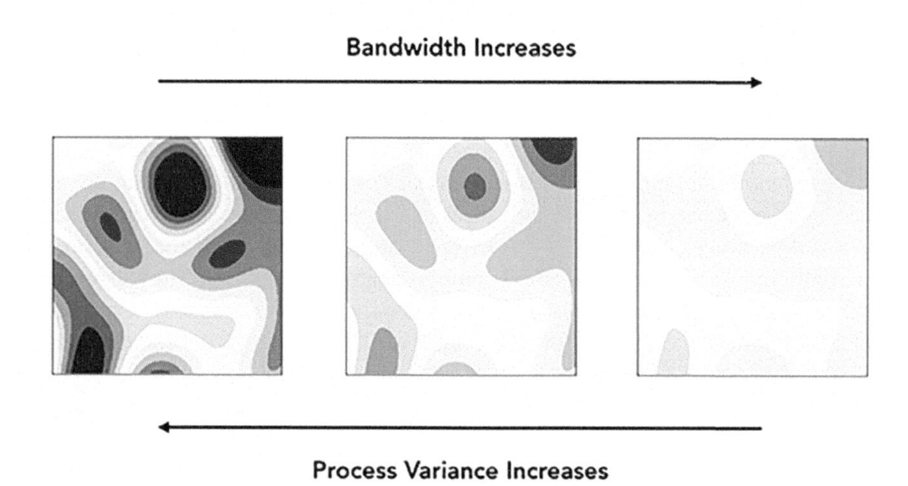

FIGURE 4.3
Relationship Between Process Variance and Bandwidth Parameter.

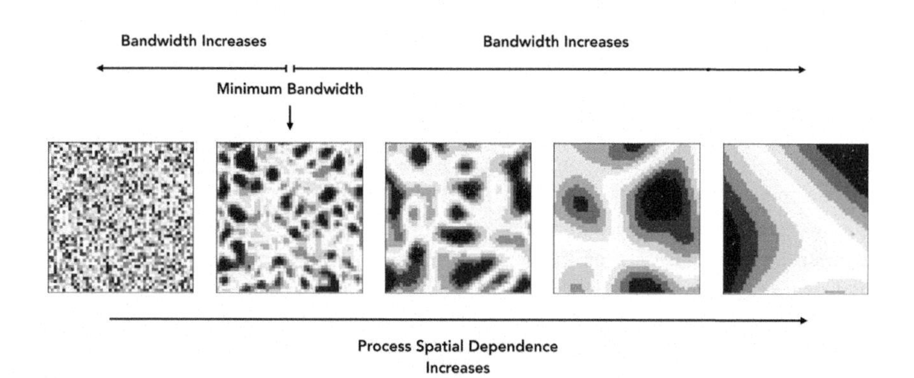

FIGURE 4.4
Relationship Between Process Dependence and Bandwidth Parameter.

spatial dependency arising from strong local structure, and then the bandwidth increases when there is either a lack of spatial dependency or extreme spatial dependency because there is little-to-no local structure. Overall, the largest bandwidths are most closely associated with the parameter surfaces that exhibit low variance while the smallest bandwidths are most closely associated with the parameter surfaces that exhibit high variance. This is because when there is little spatial variation in a process, the use of data from locations further away in the local regressions introduces little bias, and the optimal bandwidth will be large. In contrast, with fixed spatial dependency, as the process exhibits increasing spatial variation, the optimal bandwidth will decrease as using data from more distant locations will introduce more bias. These trends support the notion that small bandwidths are related to conditional associations between a covariate and the dependent variable that vary rapidly over space. In such situations, the amount of 'misinformation' in data quickly exceeds the amount of 'information' as distance from the regression point increases. When there is little or no spatial variation in the conditional associations, data can be borrowed from locations at much greater distances to reduce parameter uncertainty without much penalty as there is nothing to be gained from a small bandwidth. To summarize, small bandwidths are indicative of processes (local parameters) that vary rapidly over space, and larger bandwidths are indicative of processes that are relatively constant over space. However, this relationship between bandwidth and local process variation is partially moderated by the degree of process spatial dependency with the optimal bandwidth also tending to decrease slightly when there is higher spatial dependency (holding parameter variation constant) until very high spatial dependence when the bandwidth starts to increase again. That is, as processes initially become more spatially dependent, the inclusion of more distant locations (and more different in terms of processes) in the local calibrations would increase the bias, and therefore the optimal bandwidth decreases to avoid this. If the processes were instead approaching either a uniform or random distribution over space, this would not be an issue, and data could be borrowed from more distant locations using a larger bandwidth to reduce parameter uncertainty without increasing bias. Since spatial process variation and dependence tend to coexist, these two attributes will typically jointly determine the optimal bandwidth value with process variation being more influential.

An additional experiment from Fotheringham et al. (2022) explored the relationship between the magnitude of the model error variance and the optimal bandwidth. The experiment estimated optimal bandwidths on datasets generated based on local parameter surfaces with varying degrees of spatial dependency and levels of error variance magnitude. It was found that as the model error variance decreases, the random component of the dependent variable decreases, making the local slopes (parameters) relatively larger in comparison and therefore increasingly important in determining the dependent variable. When spatial dependence is held constant, bandwidth increases as the variance of the error term (i.e., noise) increases and process strength decreases. Because the optimized bandwidth is a trade-off between bias and variance (i.e., parameter estimate standard error), it increases to include more data in the local regression in order to reduce parameter estimate uncertainty caused by the increase in noise (greater randomness and weaker process strength).

This effect is strongest when spatial dependency is low and diminishes as spatial dependency in the parameter surface increases because high spatial dependency means only a relatively small number of data points are needed to produce parameter estimates with both low bias and variance even with higher levels of noise. The experiment also found that if the level of noise in the model is instead held constant, the optimized bandwidth decreases as spatial dependency increases. This effect is strongest for higher levels of noise and lower levels of spatial dependence, causing the optimized bandwidth to become very large and approach that of a global model. The opposite is observed when the noise in the model is consistently low such that the optimized bandwidth will tend to become increasingly smaller because adding more data will more rapidly increase bias as spatial dependence increases without substantially reducing variance.

Overall, conclusions from these experiments indicate that small bandwidths stem from processes that (i) have a high degree of spatial variability; (ii) have a high degree of spatial dependency; and (iii) are strong, in the sense that the covariate has a meaningful impact on the dependent variable. Out of these three factors, the bandwidth appears to be most sensitive to process variability though this can be moderated by low or extremely high process spatial dependence and low process strength (i.e., high noise).

4.5 Scales Issues in Local Modeling

Spatial analysis has a rich history of grappling with how to approach the appropriate scale at which phenomena should be measured, analyzed, and interpreted. And since spatial analysis involves notions of geographic scale, observation scale, and process scale, these notions can lead to problems for the spatial analyst. As attributes of spatial data, the geographic scale (or extent) and the observation scale (or size of units) are linked to two important issues. The first is perhaps more well known within spatial analysis and is often referred to as the modifiable areal unit problem (MAUP) (Openshaw, 1977; Openshaw & Taylor, 1979; Fotheringham & Wong, 1991; Jelinski & Wu, 1996; Cressie, 1996; Wong, 2004; Dark & Bram, 2007; Robinson, 1950). The MAUP describes the long-standing issue that the inferences we make based on an analysis of spatial data might vary according to the definition of the spatial units for which the data are measured. The MAUP can arise when data are aggregated to zoning schemes with either different partitions at a similar scale (e.g., stable average unit size) or similar partitions at different scales (e.g., different average unit size). Both scenarios are important and may coexist, though the latter case of different size spatial units is the primary interest here in the context of spatial scale.

The second scale problem is referred to as Simpson's paradox and describes the counterintuitive situation where aggregated data exhibit a relationship that is completely reversed when the data are disaggregated. It is relatively well known in aspatial contexts where aggregations to different non-spatial groups, such as gender or age, can yield contradictory conclusions (Cohen & Nagel, 1934; Simpson, 1951; Blyth, 1972; Bickel et al., 1975; Wilson, 2013), though it is less well known within spatial analysis where groups pertain to data at different scales (either geographic scales or observation scales). Consequently, at one spatial scale it is possible to infer that two variables are positively related, and at another spatial scale it is possible to infer they are negatively related. Both of these statements can be true, posing a substantial challenge to drawing robust inferences from the data.

A further issue relates to inferring process scales, which becomes difficult when there are processes operating at different spatial scales and therefore requires data to be collected over different geographic scales and then incorporated into a single model. In a regression context, all three of these issues are linked in that they are symptomatic of spatially varying processes.

Consider a set of processes operating across a "frame of reference" (defined by a given geographic scale) that results in the recorded measurements of variables at various locations (defined by a given observation scale) within that frame. If the processes generating the data are the same everywhere (i.e., stationary over space), then all three notions of scale described earlier (MAUP, Simpson's paradox multiscale processes) are not problematic for a regression analysis. When the set of processes operates identically at all locations, then the geographic scale or frame of reference will not be important for determining the sample of locations used to make inferences about these processes. Similarly, it also would not matter if we changed the observation scale of the units used to partition the data because every aggregation would be related to the same set of processes. This means that neither the MAUP nor Simpson's paradox is an issue in this scenario. Lastly, if none of the processes varies spatially, then this implies that the processes all operate at the same (global) scale, and there is therefore no need to differentiate and incorporate multiple process scales. Hence, the three scale issues noted earlier only exist if the processes are spatially nonstationary. In contrast, the geographic scale becomes important if processes vary over space because different extents will encapsulate data pertaining to different processes, and the observation scale becomes important because the way we bundle together individual locations produces mixes of data that pertain to different processes. In this alternative and more likely scenario, all three issues become a concern.

It has become increasingly recognized that some processes, particularly those related to human preferences, decision-making, and actions, might vary over space. Thus, when the processes being modeled are spatially nonstationary, the standard global approach to modeling whereby all the data available at a geographic scale are used to calibrate a model and produce one estimate of each parameter in the model will be prone to all three scale problems. The alternative approach of local modeling yields detailed information on spatially varying processes through the estimation of local parameter estimates and local diagnostic statistics using subsets of data at smaller geographic scales. The concept of scale is therefore implicit in any discussion of local models, and the raison d'être of local models is that a global scale (where 'global' refers to all locations at a larger predefined geographic scale) might be the incorrect scale at which to undertake any analysis of spatial processes—the alternative being a local scale (where 'local' refers to a smaller geographic scale).

Fortunately, multiscale local models, such as MGWR, allow us to incorporate multiple processes across different geographic scales, relieving one of the three scale issues. Though the issues of the MAUP and Simpson's paradox still remain, they can be better understood by moving away from the traditional view that they are a product of data properties and recasting them in terms of the properties of processes through the lens of local models (Fotheringham & Sachdeva, 2022; Sachdeva & Fotheringham, 2023).

4.5.1 The Modifiable Areal Unit Problem (MAUP)

The traditional approach to the MAUP consists of calibrating the same (global) model with the same amount of data across the same geographic scale but with the data partitioned across space in different ways and observing whether or not parameter estimates are stable. Implicit in this approach is that the processes generating the data are always assumed to be constant across the study area. An alternative approach to the MAUP focuses on processes and does not assume processes are stationary, such that a local model would be necessary to calibrate the same model with different spatial subsets of the data to reliably capture spatial processes. In this case, the calibration of a global model will be compromised because the single parameter estimate will not be indicative of the spatial variation for each process. In fact, the single parameter estimate will likely be biased compared to most of the true locally varying parameter estimates (i.e., processes), which is referred to here as process misspecification. In contrast to the traditional approach to the MAUP that focuses on attributes of the data, such as spatial autocorrelation or multicollinearity, Fotheringham and Sachdeva (2022) explore the impact of this process misspecification on the MAUP. Data were first generated from a local model that represents processes with different degrees of spatial variation, implying different levels of process misspecification when a global model is calibrated. The initial data were then aggregated to tessellations at different geographic scales, different zoning schemes while holding geographic scale constant, and a combination of these two scenarios. Finally, global models were calibrated on each of these aggregated datasets to understand the effects of the MAUP related to different levels of process spatial variation.

The results of the above experiments demonstrate that the effects of the MAUP in global models are indeed a function of the degree to which processes vary over space. For the scale aspect of the MAUP, the largest deviations in the global parameter estimates (from a global model calibrated on the initial disaggregate data) were observed for processes that had stronger true underlying spatial variation, and this was exacerbated by courser scale aggregations of the data. Similar conclusions were also obtained in regard to the zoning aspect of the MAUP. These results arise when processes are not constant over space and vary locally because different zoning systems will create mixes of data that have been produced by different processes, thus generating different parameter estimates for different zoning systems. As the processes become increasingly nonstationary, the MAUP will increase in severity. The converse situation was also explored by repeating the experiment using initial processes that were not spatially varying. When data are generated from stationary processes, the

MAUP was less evident, and there was very little difference among the global parameter estimates when using data from the different aggregations. This is because any mixing of data produces roughly the same parameter estimates, no matter what partition is employed as the processes being modeled are the same regardless of the mixing of the data. The primary insight of these results is that it is possible to view the MAUP as arising from the mixing (aggregation) of data that have been produced by locally varying processes. It is therefore especially important to use a local model (and disaggregate data) when a global model demonstrates symptoms of the MAUP.

4.5.2 Simpson's Paradox

Simpson's paradox is the reversal of a quantity of interest when the data are grouped or aggregated using different stratification schemes. There are two scale-related facets that can be used to stratify spatial data that both can lead to Simpson's paradox. The first relates to the observation scale of the data when different values of a statistic are obtained for data that have been aggregated to spatial units with a different average unit size, such as census tracts, counties, or states. When the statistical estimates diverge enough to produce qualitatively different relationships, this is a manifestation of Simpson's paradox and is essentially an extreme variant of the MAUP. The second facet relates to the geographic scale or frame of reference used to select a subset of data within the calculation of a statistic. When different frames of reference within a single study area yield qualitatively different relationships, this is also a manifestation of Simpson's paradox. Hence, Simpson's paradox is a scale issue—at one spatial scale (either observation scale or geographic scale) we might infer that $\mathbf{y}$ and $\mathbf{x}$ are positively related, but at a different spatial scale we might infer they are negatively related. Both manifestations of Simpson's paradox are important in the local modeling paradigm, though the second manifestation associated with the geographic scale used to define spatial subsets of data is particularly pertinent for local models.

The development of local models that measure relationships at different locations using subsets of data elevates the challenge of appropriate scale selection. In the traditional global modeling approach, a model is calibrated with all available data. In contrast, the alternative process-driven focus of local models involves calibrating the same model with different spatial subsets of the data and examining whether the results for each subset (local model) differ from the global model. In some cases, the contrast between local and global calibrations of the same model can result in a form of Simpson's paradox whereby the results at the local level yield the inverse inference to the results at the global level. While this may initially seem problematic, it is not an issue when the inferences from each scale are contextualized. As with aspatial examples of Simpson's paradox, statements from both the global scale and the local scale can be true because they answer different questions (or, alternatively, the same question posed at different spatial scales). Though this conclusion may seem counterintuitive, a closer look at two concrete examples originally described by Sachdeva and Fotheringham (2023) and Fotheringham and Sachdeva (2022) can help to demystify Simpson's paradox.

The first example is theoretical and focuses on crime rates and the density of vacant lots. Suppose an analyst has a set of street-level crime rates in a city as well as data on several determinants of burglary rates, one of which is the density of vacant lots. After calibrating a global model, suppose further that the conditioned relationship between burglary rates and the density of vacant lots is significantly positive. This suggests that for the entirety of this city, neighborhoods with more vacant lots are subject to more burglaries on average. If the analyst instead applies a local model, it might be found that there is a significantly negative relationship in many local neighborhoods, indicating that as the number of vacant lots increases locally, there is a decrease in opportunities for burglary. Which level of analysis is correct then—do vacant lots encourage or discourage burglaries? The answer is that both are correct; it depends upon the scale of the analysis used to pool data and measure the relationships because different processes are occurring at different scales.

A second example focuses on the relationship between house prices and their age of houses and has been demonstrated empirically in both Seattle and Los Angeles (Sachdeva & Fotheringham, 2023; Fotheringham & Sachdeva, 2022). This relationship was explored both by comparing local models to global models using individual house price data, as well as local models using data aggregated two different administrative units (block groups, census tracts, and zip codes). In both housing markets, the global hedonic price model calibration suggests that individuals prefer older housing (positive relationship), yet the calibration of a local model with the same data suggests a preference for newer housing (negative relationship). At the global scale, preferences across neighborhoods are being modeled where individuals may prefer older and perhaps more established neighborhoods. At the local scale, preferences are being modeled within neighborhoods where there might be a premium on newer housing with more modern features and fewer repairs. This suggests that decisions occur at different levels of a spatial hierarchy: at the upper level of the hierarchy, a global model captures decisions involving subsets of entities, whereas a local model captures decisions involving entities within a subset. There are two fundamentally different processes that reveal different insights about the observed phenomena at different scales, further supporting the notion that the reversal in the observed trend is inherently a scale issue.

Through a process-driven view, inferences from global and local models are not as contradictory as they appear. In addition, global estimates are not necessarily 'averages' of the associated local estimates. In some instances, a spatial variant of Simpson's paradox can arise whereby a relationship is significantly positive (negative) at the global level but is significantly positive (negative) at the local level. There is nothing wrong in this, and though it is possible for global estimates to summarize local estimates, there is no reason to expect this as the default situation. When discrepancies do arise between global and local estimates, neither perspective may be incorrect, and it is important to remember that they are potentially answering different questions related to different spatial scales. Global estimates inform on the conditioned relationship between y and x across the whole study area while local parameter estimates inform on the same conditioned relationship but around a single location. A failure to use local model, and especially one based on an optimal bandwidth parameter, leaves a spatial analysis vulnerable to missing these important scale-related differences.

4.6 Summary

This chapter is focused on the fundamental role of spatial scale within context-driven local models and explores linkages between spatial scale, spatial heterogeneity, and local models. Though there are many conceptualizations of spatial scale, typically either the geographic scale or the observation scale is varied while the other is held constant in order to learn about spatial processes. This facilitates the identification of the scale at which processes become most salient and give rise to the notion of an indicator of process scale. In particular, MGWR yields a sophisticated indicator of scale known as the bandwidth, which allows different scales to be compared, inference to be conducted, and scale-related uncertainties to be quantified. Knowledge about the bandwidth is highlighted, such as the influence of process spatial variation, spatial dependency, and process strength, and some additional scale issues are highlighted from the perspective of local modeling.

There are several important takeaways regarding scale in MGWR. First, the bandwidth is an important measure that serves as a proxy for the scale of each spatial process in the model and helps to summarize the spatial heterogeneity of each process. Second, the two issues of MAUP and Simpson's paradox can be viewed through the lens of local models and spatial processes. Both issues can be rectified through the application and interpretation of local models, especially by recognizing that different processes can be captured at different scales because of the spatial heterogeneity of processes. This insight is particularly crucial for understanding the issue of reproducibility and replicability in spatial analysis (Kedron et al., 2021a, 2021b), which is more nuanced than in other disciplines. Rather than signaling the strange and inexplicable invalidation of the analysis of spatial relationships, the MAUP and Simpson's paradox instead reflect that when we model at different scales, we are answering different questions. It is therefore imperative to judiciously apply local models, and as they become more popular, it will be increasingly important to consider how and why a process might manifest differently at different scales.

References

Aldstadt, J., & Getis, A. (2006). Using AMOEBA to create a spatial weights matrix and identify spatial clusters. *Geographical Analysis*, *38*(4), 327–343.

Atkinson, P. M., & Tate, N. J. (2000). Spatial scale problems and geostatistical solutions: A review. *The Professional Geographer*, *52*(4), 607–623.

Bickel, P. J., Hammel, E. A., & O'Connell, J. W. (1975). Sex bias in graduate admissions: Data from Berkeley. *Science*, *187*(4175), 398–404.
Blyth, C. R. (1972). On Simpson's paradox and the sure-thing principle. *Journal of the American Statistical Association*, *67*(338), 364–366.
Cohen, M. R., & Nagel, E. (1934). *An introduction to logic and scientific method*. San Diego, CA: Harcourt, Brace.
Cressie, N. (1996). Change of support and the modifiable areal unit problem. *Faculty of Informatics—Papers (Archive)*, 159–180.
Csillag, F., Fortin, M.-J., & Dungan, J. L. (2000). On the limits and extensions of the definition of scale. *Bulletin of the Ecological Society of America*, *81*(3), 230–232.
Dabiri, Z., & Blaschke, T. (2019). Scale matters: A survey of the concepts of scale used in spatial disciplines. *European Journal of Remote Sensing*, *52*(1), 419–434.
Dark, S. J., & Bram, D. (2007). The modifiable areal unit problem (MAUP) in physical geography. *Progress in Physical Geography: Earth and Environment*, *31*(5), 471–479.
Dungan, J. L., Perry, J. N., Dale, M. R. T., Legendre, P., Citron-Pousty, S., Fortin, M.-J., Jakomulska, A., Miriti, M., & Rosenberg, M. S. (2002). A balanced view of scale in spatial statistical analysis. *Ecography*, *25*(5), 626–640.
Fotheringham, A. S., Brunsdon, C., & Charlton, M. E. (2002). *Geographically weighted regression: The analysis of spatially varying relationships*. London: Wiley.
Fotheringham, A. S., & Sachdeva, M. (2022). Scale and local modeling: New perspectives on the modifiable areal unit problem and Simpson's paradox. *Journal of Geographical Systems*, *24*, 475–499.
Fotheringham, A. S., & Wong, D. W. S. (1991). The modifiable areal unit problem in multivariate statistical analyses. *Environment and Planning A*, *23*, 1025–1044.
Fotheringham, A. S., Yu, H., Wolf, L. J., Oshan, T. M., & Li, Z. (2022). On the notion of 'bandwidth' in geographically weighted regression models of spatially varying processes. *International Journal of Geographical Information Science*, *36*(8), 1485–1502.
Getis, A., & Aldstadt, J. (2004). Constructing the spatial weights matrix using a local statistic. *Geographical Analysis*, *36*(2), 90–104.
Gibson, C. C., Ostrom, E., & Ahn, T. K. (2000). The concept of scale and the human dimensions of global change: A survey. *Ecological Economics*, *32*(2), 217–239.
Goodchild, M. F. (2001). Metrics of scale in remote sensing and GIS. *International Journal of Applied Earth Observation and Geoinformation*, *3*(2), 114–120.
Jelinski, D. E., & Wu, J. (1996). The modifiable areal unit problem and implications for landscape ecology. *Landscape Ecology*, *11*(3), 129–140.
Jenerette, D., & Wu, J. (2000). On the definitions of scale. *Bulletin of the Ecological Society of America*, *81*(1), 104–105.
Kedron, P., Frazier, A. E., Trgovac, A. B., Nelson, T., & Fotheringham, A. S. (2021b). Reproducibility and replicability in geographical analysis. *Geographical Analysis*, *53*(1), 135–147.
Kedron, P., Li, W., Fotheringham, A. S., & Goodchild, M. F. (2021a). Reproducibility and replicability: Opportunities and challenges for geospatial research. *International Journal of Geographical Information Science*, *35*(3), 427–445.
Lam, N. S.-N., & Quattrochi, D. A. (1992). On the issues of scale, resolution, and fractal analysis in the mapping sciences. *The Professional Geographer*, *44*(1), 88–98.
Manson, S. M. (2008). Does scale exist? An epistemological scale continuum for complex human—environment systems. *Geoforum*, *39*(2), 776–788.
Marceau, D. J. (1999). The scale issue in social and natural sciences. *Canadian Journal of Remote Sensing*, *25*(4), 18.

Marceau, D. J., & Hay, G. J. (1999). Remote sensing contributions to the scale issue. *Canadian Journal of Remote Sensing, 25*(4), 357–366.

Meentemeyer, V. (1989). Geographical perspectives of space, time, and scale. *Landscape Ecology, 3*(3–4), 163–173.

Miller, H. J. (2004). Tobler's first law and spatial analysis. *Annals of the Association of American Geographers, 94*(2), 284–289.

Openshaw, S. (1977). A geographical solution to scale and aggregation problems in region-building, partitioning and spatial modelling. *Transactions of the Institute of British Geographers, 2*(4), 459–472.

Openshaw, S., & Taylor, P. J. (1979). A million or so correlation coefficients: Three experiments on the modifiable areal unit problem. In N. Wrigley (Ed.), *Statistical applications in the spatial sciences* (pp. 127–144). London: Pion.

Oshan, T. M., Wolf, L. J., Sachdeva, M., Bardin, S., & Fotheringham, A. S. (2022). A scoping review on the multiplicity of scale in spatial analysis. *Journal of Geographical Systems, 24*, 293–324.

Robinson, W. S. (1950). Ecological correlations and the behavior of individuals. *American Sociological Review, 15*(3), 351–357.

Rogerson, P. A. (2015). Maximum Getis—Ord statistic adjusted for spatially autocorrelated data. *Geographical Analysis, 47*(1), 20–33.

Rogerson, P. A. (2021). *Spatial statistical methods for geography*. London: SAGE.

Rogerson, P. A., & Kedron, P. (2012). Optimal weights for focused tests of clustering using the local Moran statistic. *Geographical Analysis, 44*(2), 121–133.

Ruddell, D., & Wentz, E. A. (2009). Multi-tasking: Scale in geography. *Geography Compass, 3*(2), 681–697.

Sachdeva, M., & Fotheringham, A. S. (2023). A geographical perspective on Simpson's paradox. *Journal of Spatial Information Science, 26*, 1–25.

Silbernagel, J. (1997). Scale perception: From cartography to ecology. *Bulletin of the Ecological Society of America, 78*(2), 166–169.

Simpson, E. H. (1951). The interpretation of interaction in contingency tables. *Journal of the Royal Statistical Society. Series B (Methodological), 13*(2), 238–241.

Tobler, W. R. (1970). A computer movie simulating urban growth in the Detroit region. *Economic Geography, 46*, 234–240.

Wilson, R. (2013). Changing geographic units and the analytical consequences: An example of Simpson's paradox. *Cityscape, 15*(2), 289–304.

Wong, D. W-S. (2004). The modifiable areal unit problem (MAUP). In D. G. Janelle, B. Warf, & K. Hansen (Eds.), *WorldMinds: Geographical perspectives on 100 problems: Commemorating the 100th anniversary of the Association of American Geographers 1904–2004* (pp. 571–575). New York: Springer.

Wu, H., & Li, Z.-L. (2009). Scale issues in remote sensing: A review on analysis, processing and modeling. *Sensors, 9*(3), 1768–1793.

Wu, J., & Li, H. (2006). Concepts of scale and scaling. In *Scaling and uncertainty analysis in ecology* (p. 13). New York: Springer.

Yu, H., Fotheringham, A. S., Li, Z., Oshan, T. M., & Wolf, L. J. (2020). On the measurement of bias in geographically weighted regression models. *Spatial Statistics, 38*, 100453.

5

Software for MGWR

5.1 Introduction to MGWR 2.2

Having elaborated on why, and in what situations, local models are useful and described both the mechanics of MGWR and inference for MGWR in previous chapters, we now discuss computational aspects of MGWR and demonstrate the use of specialized software termed 'MGWR 2.2'. This chapter hence serves as a reference for users of the software as well as describing the computational aspects of a typical MGWR calibration. The earliest version of the MGWR software (version 1.0) was launched in 2018 with a user-friendly graphic interface that wraps around the back-end codebase of the open-source python library *mgwr* (Oshan et al., 2019). The most current version, MGWR 2.2, was launched in 2020 and integrates recent statistical and computational improvements and feature enhancements (Fotheringham et al., 2017; Fotheringham & Oshan, 2016; Oshan et al., 2019; Yu et al., 2020; Li & Fotheringham, 2020; Li et al., 2020). The software is available for both Windows 10 and MacOS systems. The back-end python package can be directly called under Windows and Unix-based systems, and the usage of it is described in Section 5.9. MGWR 2.2 is freely downloadable from the Spatial Analysis Research Center (SPARC), Arizona State University at https://sgsup.asu.edu/sparc/mgwr, and the mgwr python package is hosted at https://github.com/pysal/mgwr. Figure 5.1 shows an overview of MGWR 2.2, including data input, model specification, model calibration, and the two forms of model outputs. The detailed use of the software is described in the subsequent sections.

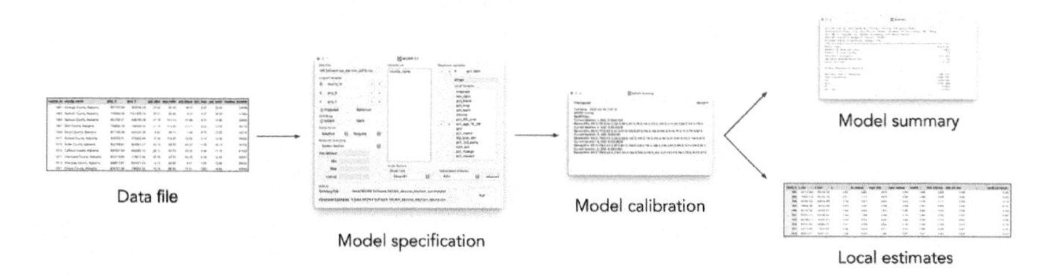

FIGURE 5.1
MGWR 2.2 Overview.

DOI: 10.1201/9781003435464-5

5.2 Graphical Interface

Once launched, MGWR 2.2 features a simple and user-friendly interface, as shown in Figure 5.2. The interface is made up of 11 major components:

1. *Data Files*: Users can browse a directory to load data. The required data file is discussed in Section 5.2.1.
2. *Variable List*: After successfully loading the data, all fields will appear in the *Variable List*, and users can use the left and right arrow buttons to specify location and regression variables for the model calibration.
3. *Location Variables*: Users must define the location variables (those containing the coordinates of the locations for which data are recorded) as well as whether the coordinates are projected (e.g., UTM) or spherical (e.g., longitude and latitude).
4. *Regression Variables*: The dependent variable and covariates must be specified by the user. If the model is a Poisson model, the user can include an offset variable.

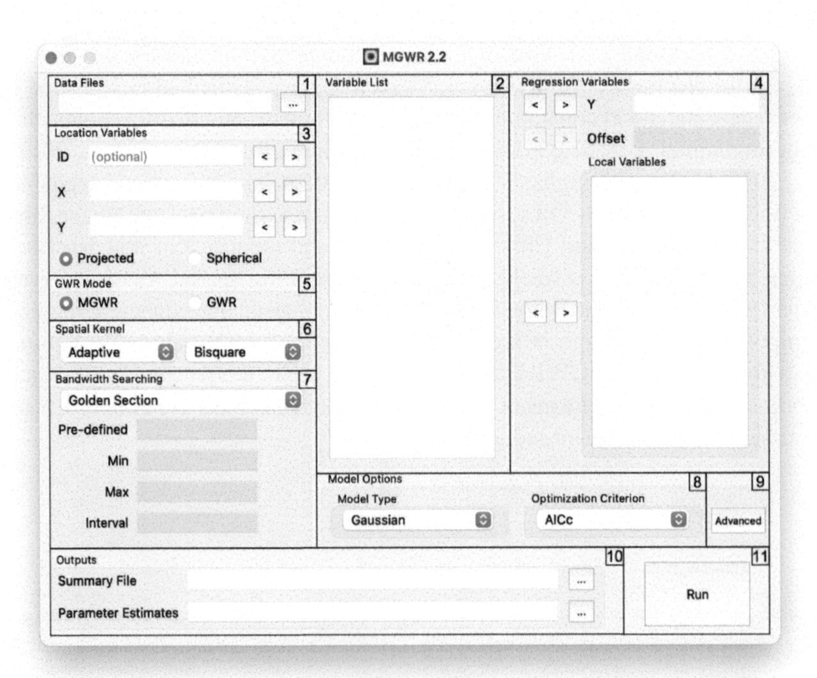

FIGURE 5.2
MGWR 2.2 Graphical User Interface.

5. *GWR Mode*: Users have the option of fitting an MGWR model (the default) or a GWR model.
6. *Spatial Kernel*: Users select the type of spatial kernel to be employed in the calibration (adaptive/fixed) as well as the kernel function (bi-square/Gaussian).
7. *Bandwidth Searching*: Users can use predefined bandwidths, an interval search, or a golden-section search routine to find optimal bandwidths.
8. *Model Options*: Users can choose from fitting Gaussian, Binomial, and Poisson models, and select the bandwidth optimization criterion (*AICc/AIC/BIC/CV*).
9. *Advanced Options*: Users can enable advanced options as discussed later.
10. *Outputs*: Users need to select a local path to store both the model summary and the location-specific model results.
11. *Run*: Clicking the button starts the model fitting process.

5.3 Data Files

This section describes the structure of the data needed to run MGWR 2.2. The data need to be in a tabular file format of excel (*.xls, *.xlsx), comma delimited (*.csv), or dbase IV (*.dbf). The data file has to contain the variable names as the first row, and data records as the subsequent rows (one row for each location). Note that any row with missing or non-numerical data will be dropped from the analysis. A data file usually contains the following fields to fit a GWR or MGWR model:

- One column to uniquely identify each observation (optional).
- Two columns containing the X and Y coordinates of the location at which data are recorded.
- One column indicating the dependent variable.
- One or more columns indicating the covariates.

If data relate to areas, some representative location of that area, such as the geometric or population-weighted centroid of the area, should be used to provide the relevant X and Y coordinates. Regardless of whether the data pertain to points or areas, we refer to these observations as being the data points. As an example, Figure 5.3 shows some of the data on voting in the 2016 US Presidential election with associated socioeconomic attributes at the county level that will be used in Section 5.8 to demonstrate the use of the software.

ID Variables (optional) | Location Variables | Dependent Variable | Covariates

county_id	county_name	proj_X	proj_Y	pct_dem	sex_ratio	pct_black	pct_hisp	pct_bach	median_income
1001	Autauga County, Alabama	837137.90	-833746.20	24.62	95.40	18.37	2.57	24.60	53099
1003	Baldwin County, Alabama	759093.38	-1051545.14	20.21	95.30	9.23	4.37	29.50	51365
1005	Barbour County, Alabama	957282.21	-898765.29	47.18	115.10	47.89	4.31	12.90	33956
1007	Bibb County, Alabama	788866.38	-784264.43	21.76	115.20	21.21	2.22	12.00	39776
1009	Blount County, Alabama	827160.38	-665520.38	8.62	98.10	1.56	8.73	13.00	46212
1011	Bullock County, Alabama	925253.41	-875362.69	75.59	118.30	75.50	0.12	10.30	29335
1013	Butler County, Alabama	842709.91	-924527.21	43.19	88.50	43.52	1.25	16.10	34315
1015	Calhoun County, Alabama	894531.69	-683083.72	28.72	92.70	20.33	3.44	17.70	41954
1017	Chambers County, Alabama	943719.96	-778213.84	42.45	92.00	40.48	0.44	12.40	36027
1019	Cherokee County, Alabama	908816.87	-634071.24	14.73	99.00	4.61	1.56	13.90	38925
1021	Chilton County, Alabama	826757.96	-798204.80	16.18	96.90	10.01	7.62	14.80	42594

FIGURE 5.3
An Example Dataset for MGWR 2.2.

Once the data are successfully loaded into MGWR 2.2, the user needs to specify the location variables, the dependent variable, and the covariates to be used in the model by moving variables (>) from the list into the relevant sections of the GUI. If the coordinates are latitudes and longitudes, the user needs to select the *spherical* radio button.

5.4 Spatial Kernels

Kernels are used to define the local weighting scheme in GWR or MGWR. Users have the option to choose a kernel function and a kernel type from the 'Spatial Kernel' dropdown lists. The currently available kernels are adaptive bi-square and fixed Gaussian, the formulas for which are shown in equations (5.1) and (5.2). The adaptive bi-square kernel fits local regression using nearest neighbors, and local bi-square weights decrease to zero at the Nth nearest neighbor. The fixed Gaussian kernel employs a distance-based bandwidth, and the local weights do not go down to zero but at the reported bandwidth (which is 2.45 times the Gaussian kernel formula bandwidth) the weights fall below 0.05 and can be discounted. A similar practice was employed in a geostatistical context by Cressie (1993). Essentially, the degree of distance-decay in the bi-square and Gaussian kernels is very similar, as shown in Figure 5.4. An adaptive kernel is preferred in most cases, especially in situations where the density of locations varies over space because in such situations the use of a fixed kernel may mean that some of the local regressions will be based on unacceptably small numbers of data points.

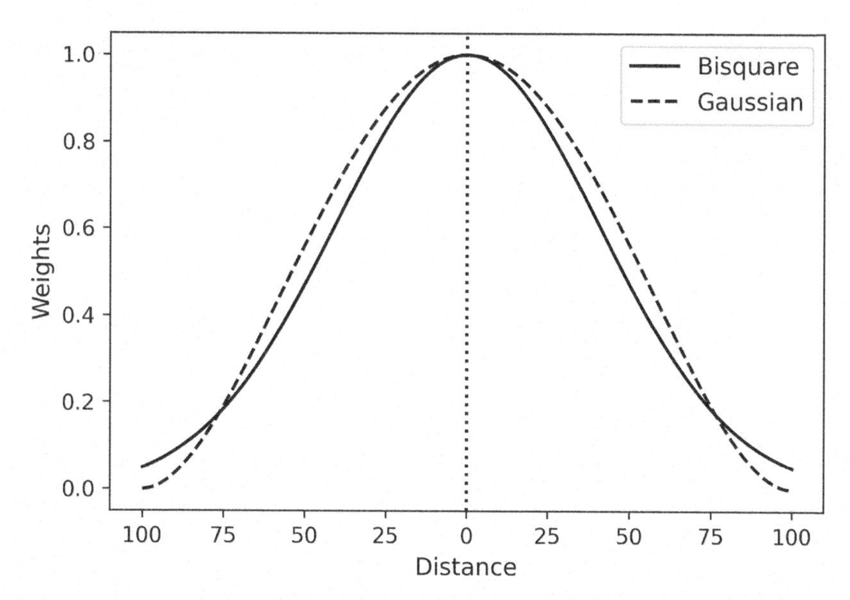

FIGURE 5.4
Bi-square and Gaussian Kernel Distance-Decay Weighting Functions.

Gaussian $$w = exp\left(-\frac{1}{2}\left(c*\left(\frac{d}{b}\right)\right)\right)^2 \qquad c = 2.45 \tag{5.1}$$

Bi-square $$w = \left(1 - \left(\frac{d}{b}\right)^2\right)^2 \; if\; d < b\,0\,otherwise \tag{5.2}$$

5.5 Bandwidth Selection

Bandwidth selection is carried out either by optimizing a model fit criterion or by manual specification. Optimal selection is preferred when, as is usual, there is no theoretical guide to manually specify the bandwidth. To use a predefined bandwidth, users need to input one single numerical bandwidth for GWR (e.g., 100) or a list of covariate-specific bandwidths (e.g., [100, 150, 50, 300] for a model with four covariates, including the intercept) for MGWR. In most circumstances, no prior knowledge of the bandwidth is available, and then users should adopt a bandwidth-searching method to optimize a specific model selection criterion. The supported objective functions are *AIC*, *AICc*, *BIC*, and *CV* (see Chapter 2 for details). The default option is *AICc*, which penalizes smaller bandwidths that result in more complex models

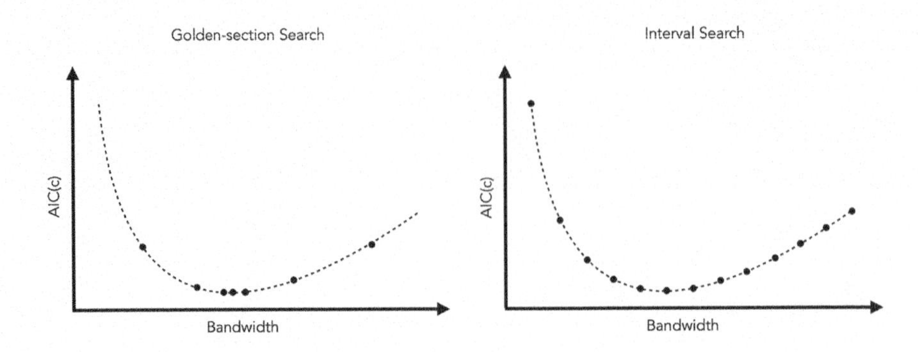

FIGURE 5.5
Golden-Section (Left) and Interval (Right) Routines to Search for Optimal Bandwidths.

and which consume more degrees of freedom. The bandwidth-searching routine can be either interval search where the user needs to specify the bounds and the step of the search, or golden-section search, which is a 1-D heuristic optimization method that subsequently checks golden-section points to narrow the search range. The two search routines are shown in Figure 5.5 in which the black dots indicate a history of bandwidths at which the *AICc* function is evaluated.

5.6 Advanced Settings

Seven advanced options can be selected by clicking the 'Advanced' button, as shown in Figure 5.6.

- *Variable Standardization*: This option performs a z-transformation on the dependent variable and covariates so that each variable has a mean of 0 and a standard deviation of 1. This option is available for both GWR and MGWR models. The default setting is "*on*", and for MGWR this is recommended to always leave it on, as doing so will allow for a scale-free interpretation and comparison of the covariate-specific bandwidths.
- *Initialization*: Users have the option to initialize the MGWR backfitting process (for details, see Chapter 2) with either a GWR model or an OLS model. The final model results should not be sensitive to the initialization method, but the convergence rate may vary depending on the complexity of the underlying relationships in the data. The default setting is that the MGWR calibration routine uses the GWR-derived local parameter estimates as a starting point.
- *Measure of Score of Change (SOC)*: Users may change the measure of score of change, which is the termination criterion for the backfitting iterations

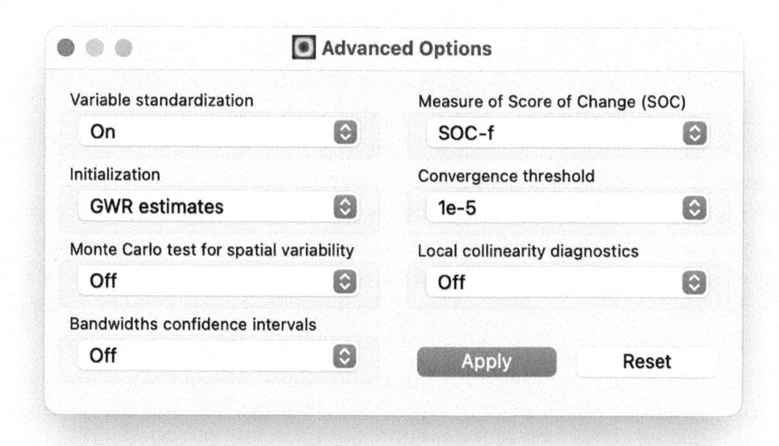

FIGURE 5.6
Advanced Options User Interface.

to be deemed to have converged. Two types of SOC measures can be used: (i) *SOC-RSS*, which is the proportional change in model's residual sum of squares (RSS) between current and previous iterations (equation 5.3), and (ii) *SOC-f*, which is the change in the GWR smoother (f) (equation 5.4). Both options are scale-free, but the *SOC-f* option has the advantage of focusing on relative changes in the additive terms rather than on the overall model fit. However, it also may take longer to converge in some cases. The *SOC-f* option is the default setting.

$$SOC_{RSS} = \frac{\left|RSS_{new} - RSS_{old}\right|}{RSS_{new}} \tag{5.3}$$

$$SOC_f = \sqrt{\frac{\sum_{j=1}^{k} \frac{\sum_{i=1}^{m}\left(\hat{f}_{ij}^{new} - \hat{f}_{ij}^{old}\right)^2}{m}}{\sum_{i=1}^{m}\left(\sum_{j=1}^{k}\hat{f}_{ij}^{new}\right)^2}} \tag{5.4}$$

- *Convergence Threshold*: This is only available for MGWR models. Users may change the convergence threshold (10^{-3} or 10^{-5}), which determines the threshold of change below which the model is deemed to have converged. The default is 10^{-5}.

- *Monte Carlo Test for Spatial Variability*: To test whether the spatial variability of the local estimates is attributable to sampling variation versus a result of more meaningful process nonstationarity (see Chapter 3 for details), MGWR 2.2 provides the option to run a Monte Carlo test for both GWR and MGWR local parameter estimates. The Monte Carlo test runs once to derive the local parameter estimates and then repeatedly derives new local parameter estimates after randomly rearranging the data points to measure whether the variability of each parameter surface derived from the real data could have arisen by chance. There are 1,000 realizations for the Monte Carlo test, and therefore it is computationally intensive and may significantly increase the model run time. The default setting is "off", and we recommend only turning this option on once everything else about the model calibration has been checked.
- *Local collinearity diagnostics*: Local collinearity diagnostics will return a local condition index called a '*local_CN*', which identifies the number of near dependencies among the columns of the design matrix (see Chapters 2 and 7 for more details on multicollinearity in local models). In addition, the local collinearity diagnostics also provide a variance decomposition proportion ('*local_VDP*') and a local variance inflation factor ('*local_VIF*') for each covariate, which, in conjunction with the condition index, provide a measure of the degree to which the corresponding regression estimate has been degraded by the presence of multicollinearity.[1] Additional information can be found in Fotheringham and Oshan (2016), and also the reader is referred to Section 5.7.4. The default setting is '*off*'.
- *Bandwidth Confidence Interval*: When turned on, the bandwidth confidence interval will provide the lower and upper bounds corresponding to the 95% confidence interval for each bandwidth in either an MGWR or GWR calibration. The bandwidths' confidence intervals are computed based on the Akaike weights procedure developed in Li et al. (2020), and more details can be found in Chapter 3. The default setting is '*off*'.

5.7 Local Diagnostics

In contrast to traditional global modeling in which parameters and diagnostics are constant over space, local models such as MGWR produce location-specific parameter estimates and diagnostics. There are four groups of location-specific outputs from the MGWR 2.2 software:

1. Input data information, which includes ID, coordinates, and the original dependent variable;
2. Standard regression outputs, which include the local model residuals and predicted dependent variable from GWR or MGWR as well as the global

model residuals from OLS or GLS. Residuals from both global and local models can be used to compute spatial autocorrelation statistics to infer whether the local modeling approach removes the dependency often reported in the residuals from global models (see Chapter 6);

3. Local diagnostics, which include the local R^2, local parameter estimates, local t-values, local p-values, and the sum of the weights used in each local regression calibration; and
4. Local collinearity measures (if turned on in the 'Advanced' settings—see Section 5.6).

The first two groups of output are standard and self-explanatory; in the following we describe the local diagnostics and collinearity measures further.

5.7.1 Local R^2

The R^2 value is a common statistic to evaluate the goodness-of-fit of a regression model. In contrast to a global model in which a single R^2 value is calculated, local models can generate location-specific R^2 values, which provide a sense of how the model performs locally. The calculation of the local R^2 for a GWR model is as follows:

$$R_i^2 = \left(TSS_i^w - RSS_i^w\right) / TSS_i^w \tag{5.5}$$

$$TSS_i^w = \sum_j w_{ij} \left(y_j - \bar{y}\right)^2 \tag{5.6}$$

$$RSS_i^w = \sum_j w_{ij} \left(y_j - \hat{y}_j\right)^2 \tag{5.7}$$

where TSS_i^w is the geographically weighted total sum of squares and RSS_i^w is the geographically weighted residual sum of squares, and w_{ij} describes the weight of data point j in the local regression calibrated at location i. In a GWR model, the weight matrix $\boldsymbol{W}$ is calculated based on the optimal bandwidth for each location. However, in MGWR, due to the backfitting estimation and multiple bandwidths, no single weight matrix is available. In this case, MGWR 2.2 uses the bandwidth and weight matrix from a GWR model to compute equations (5.6) and (5.7) and so produces a set of "pseudo" local R_i^2 values.

5.7.2 Local Estimates and Inference

Local parameter estimates are included in the output file together with associated standard errors. The estimates and standard errors then are used to compute the local t-values, and a local p-value can be computed based on the t-distribution and the associated degree of freedom.

$$t_{ij} = \frac{\widehat{\beta}_{ij}}{se_{ij}} \tag{5.8}$$

Then, following the traditional *t*-test approach in standard regression, users can make inferences about the significance of the local parameter estimates. It is worth noting that in GWR and MGWR, a multiple hypothesis testing issue arises (see Chapter 3 for details), and also the tests are not independent due to the data-borrowing scheme in GWR/MGWR, which makes the local parameter estimates spatially correlated. To account for both of these issues, a Bonferroni corrected *p*-value (da Silva & Fotheringham, 2016) that is based on the effective number of parameters needs to be calculated as shown here:

$$\alpha = \frac{\alpha^*}{\frac{ENP}{k}} \tag{5.9}$$

where α^* is the desired type I error rate (for example, 0.05), *ENP* is the effective number of parameters, and *k* is the number of covariates. Note that for an MGWR model, the corrected alpha needs to be covariate specific, and the equation is shown in (5.10). Then, a critical adjusted *t*-value can be calculated based on the corrected *p*-value and *t*-distribution.

$$\alpha_j = \frac{\alpha^*}{ENP_j} \tag{5.10}$$

5.7.3 Local Sum of Weights

Each local calibration uses a unique set of weighted data—the data being weighted by their proximity to the calibration focal point. Consequently, there is no guarantee that the sum of the weighted data will be a constant across the local calibrations—indeed, this would only happen if all the processes being modeled were spatially homogeneous. In order to get a feel for the actual amount of data used in each local regression, the sum of weights (equation [5.11]) is computed for each location. For processes that are global, all locations should use all the data so the sum of weights would equal *m*. For processes that are local, only a subset of data will be used in each local calibration, and the sum of weights, W_i, will reflect the effective number of data points used locally.

$$W_i = \sum_{j}^{m} w_{ij} \tag{5.11}$$

5.7.4 Local Collinearity Diagnostics

If the collinearity diagnostics option is turned on in the 'Advanced' settings, the output file will include local variance inflation factors (*VIF*) (GWR only), local condition numbers (*CN*), and local variance decomposition proportions (*VDP*). Due to the fact that MGWR has multiple spatial weight matrices, the computation of *VIF* is not as straightforward as in a GWR model, so *VIF* is not currently available for MGWR. For local *CN* and local *VDP*, which do not require a single weight matrix, these are calculated for the MGWR model. The formulas for the local collinearity statistics can be found in Chapter 2. Large *VDP* and *CN* values indicate a potential issue of collinearity, and, following Belsley et al. (2005), a *CN* greater than 10 and pairwise variance decomposition factors greater than 0.5 call for caution in interpreting the output and for a possible reexamination of the global model.

5.8 An Empirical Example of the Use of MGWR 2.2

5.8.1 Example Dataset

In this section, we present an empirical study of county-level voter preferences in the 2016 US Presidential election based on data described by Fotheringham et al. (2021). We will revisit this problem in greater detail for the 2020 election in Chapter 7 where the focus is on interpreting the local model calibration results to inform on voting behavior in the 2020 election. Here we do not interpret the results in terms of voting behavior but merely use these data to demonstrate the use of the MGWR software and how to interpret the results. The data were obtained from the MIT Election Lab and the American Community Survey 2012–2016 five-year estimates. Counties with fewer than 5,000 population were removed from the analysis to minimize the bias that would occur from using percentages based on small populations. The variables and their descriptions are given in Table 5.1.

5.8.2 Specifying the Model

The first step is to open a data file in a supported format from a local directory. This is done by pressing the [. . .] symbol and searching your local directories for the file as shown in Figure 5.7 (a). After loading the data, the headings of each column in the file will appear in the variable list, and then the arrow buttons, [>] and [<], can be used to add or remove variables into or from various sections on the model editor as shown in Figure 5.7 (b). After specifying the relevant independent and dependent variables and selecting the model settings as shown in Figure 5.7 (c), the output files are named to save the model summary text file (*_summary.txt) and the local results table (*_results.csv) in the Outputs area of the GUI as shown in Figure 5.7 (d).

5.8.3 Running the Model

After completing all the required fields, click the [Run] button as shown in Figure 5.8 (a) to begin the model calibration process. A pop-up window will appear with a

TABLE 5.1

Variables Used in the Empirical Example.

Variable	Description
Sex_ratio	The ratio of males to females with 100 representing an equal number of males and females
Pct_age_18_29	Percentage of the population aged 18 to 29
Pct_age_65	Percentage of the population aged 65 and over
Pct_Black	Percentage of the population identifying as Black
Pct_Hispanic	Percentage of the population identifying as Hispanic
Median_income	Median household income
Pct_Bachelor	Percentage of the population with a bachelor's degree or higher
Gini	Gini index of income inequality; the greater the index value, the greater the income disparity
Pct_Manuf	Percentage of the population employed in the manufacturing sector
Ln(pop_den)	Natural logarithm of population density (persons per square mile). This acts as an urban/rural indicator
Pct_3rd_Party	Percentage of votes that went to a third-party candidate
Turnout	Voter turnout—the percentage of those eligible to vote who actually voted
Pct_FB	Percentage of the population identifying as being born outside the United States
Pct_Uninsured	Percentage of population without health insurance coverage

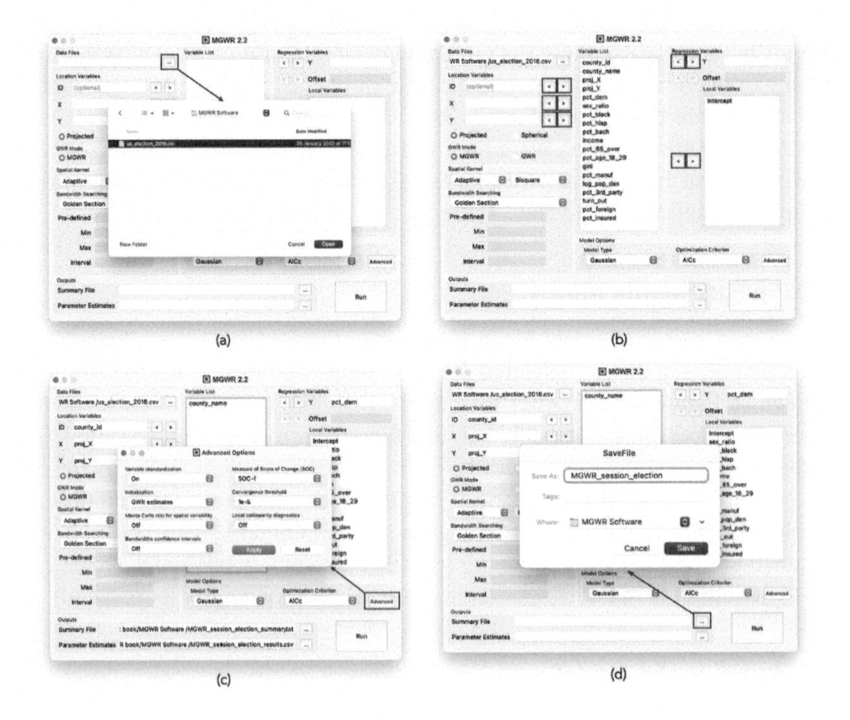

FIGURE 5.7

User Interface Demonstrating How to Load Data, Specify a Model, and Save the Outputs.

timer, which allows monitoring of the model calibration progress. An alert window with execution time will appear once the calibration is completed, as shown in Figure 5.8 (b). On clicking [OK] the model summary file will appear (Figure 5.8 [c]), and the model summary and the local model results are saved to the designated files.

5.8.4 Model Summary

As shown in Figure 5.9, the model summary has three main parts. The first part describes the general model setting, including the model type, number of observations, number of covariates, name of the dependent variable, whether the variables are standardized, and the total run time. The second part shows the global model results from an OLS model (if modeling a Gaussian response) or a GLS model (if modeling a binomial or Poisson response). Overall model statistics such as residual sum of squares, log-likelihood, AIC, $AICc$, R^2, and adjusted R^2 values are given. The

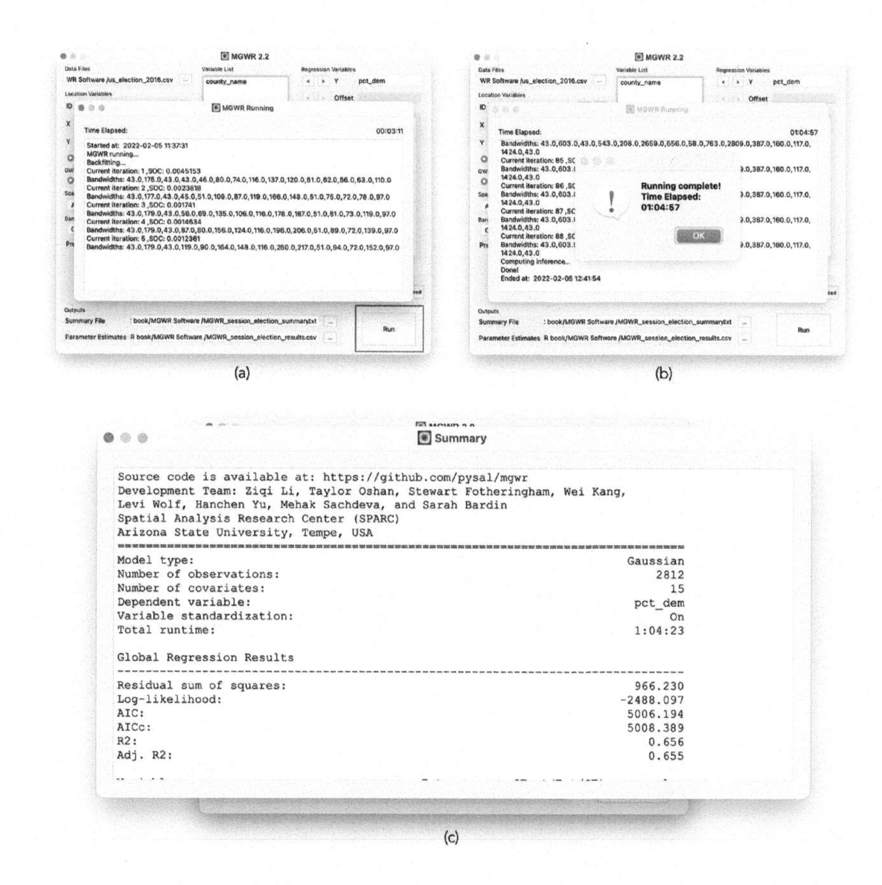

FIGURE 5.8
User Interface Showing the Model Fitting Process.

```
Part 1: General Model Settings
===========================================================================
Model type:                                                        Gaussian
Number of observations:                                                2812
Number of covariates:                                                    15
Dependent variable:                                                 pct_dem
Variable standardization:                                                On
Total runtime:                                                      1:04:23

Part 2: Global Regression Results
---------------------------------------------------------------------------
Residual sum of squares:                                            966.230
Log-likelihood:                                                   -2488.097
AIC:                                                               5006.194
AICc:                                                              5008.389
R2:                                                                   0.656
Adj. R2:                                                              0.655

Variable                                  Est.         SE  t(Est/SE)    p-value
------------------------------------ ---------- ---------- ---------- ----------
Intercept                                0.000      0.011      0.000      1.000
sex_ratio                                0.010      0.013      0.808      0.419
pct_black                                0.528      0.015     35.974      0.000
pct_hisp                                 0.283      0.019     15.005      0.000
pct_bach                                 0.425      0.026     16.530      0.000
income                                  -0.310      0.025    -12.333      0.000
pct_65_over                             -0.014      0.019     -0.737      0.461
pct_age_18_29                           -0.021      0.019     -1.094      0.274
gini                                     0.027      0.016      1.669      0.095
pct_manuf                                0.019      0.014      1.426      0.154
log_pop_den                              0.166      0.017      9.811      0.000
pct_3rd_party                            0.163      0.014     11.376      0.000
turn_out                                 0.168      0.019      8.853      0.000
pct_foreign                              0.192      0.021      9.322      0.000
pct_insured                              0.174      0.016     10.692      0.000

Part 3: Multiscale Geographically Weighted Regression (MGWR) Results
---------------------------------------------------------------------------
Coordinates type:                                                 Projected
Spatial kernel:                                           Adaptive bisquare
Criterion for optimal bandwidth:                                       AICc
Score of change (SOC) type:                                     Smoothing f
Termination criterion for MGWR:                                     1.0e-05
Number of iterations used:                                               88

MGWR bandwidths
---------------------------------------------------------------------------
Variable             Bandwidth      ENP_j   Adj t-val(95%)              D_j
Intercept               43.000    118.084            3.529            0.958
sex_ratio              603.000      9.401            2.789            0.997
pct_black               43.000    118.601            3.530            0.958
pct_hisp               543.000      6.013            2.641            0.998
pct_bach               208.000     21.019            3.041            0.993
income                2659.000      1.206            2.040            1.000
pct_65_over            656.000      6.397            2.662            0.998
pct_age_18_29           58.000    104.952            3.498            0.963
gini                   763.000      6.754            2.680            0.998
pct_manuf             2809.000      1.137            2.015            1.000
log_pop_den            387.000     10.599            2.828            0.997
pct_3rd_party          160.000     21.558            3.049            0.993
turn_out               117.000     39.958            3.230            0.986
pct_foreign           1424.000      2.808            2.371            0.999
pct_insured             43.000    132.897            3.561            0.953

Diagnostic Information
---------------------------------------------------------------------------
Residual sum of squares:                                            138.347
Effective number of parameters (trace(S)):                          601.386
Degree of freedom (n - trace(S)):                                  2210.614
Sigma estimate:                                                       0.250
Log-likelihood:                                                     244.660
Degree of Dependency (D):                                             0.986
Degree of Dependency (D*):                                            0.535
AIC:                                                                715.452
AICc:                                                              1044.591
BIC:                                                               4294.618
R2:                                                                   0.951
Adj. R2:                                                              0.937

Summary Statistics For MGWR Parameter Estimates
---------------------------------------------------------------------------
Variable                   Mean        STD        Min     Median        Max
-------------------- ---------- ---------- ---------- ---------- ----------
Intercept                -0.017      0.449     -0.980      0.003      0.894
sex_ratio                -0.029      0.028     -0.080     -0.026      0.027
pct_black                 0.723      0.220     -0.094      0.753      1.361
pct_hisp                  0.290      0.100      0.161      0.258      0.491
pct_bach                  0.354      0.179      0.132      0.286      0.831
income                   -0.207      0.004     -0.220     -0.205     -0.204
pct_65_over               0.007      0.040     -0.106      0.014      0.058
pct_age_18_29            -0.006      0.109     -0.401      0.018      0.284
gini                     -0.013      0.035     -0.067     -0.015      0.068
pct_manuf                -0.026      0.002     -0.031     -0.025     -0.025
log_pop_den               0.094      0.095     -0.087      0.106      0.258
pct_3rd_party             0.310      0.196     -0.081      0.335      0.603
turn_out                  0.037      0.084     -0.292      0.040      0.280
pct_foreign               0.047      0.045     -0.031      0.054      0.106
pct_insured              -0.032      0.143     -0.833     -0.013      0.318
===========================================================================
```

FIGURE 5.9
MGWR Summary File: General Model Setting, Global Model Results, and Local Model Summary.

summary file also contains global parameter estimates, standard errors, *t*-values, and *p*-values relating to the covariates in the global model.

The third part contains the MGWR model fitting results. The model settings are listed, including the coordinate type, kernel settings, optimization criterion, change type score, termination threshold, and the total number of iterations needed before the calibration converges. Model goodness-of-fit measures are reported and can be compared to the equivalent diagnostics for the global model. The effective number of parameters (*ENP*) is calculated as the trace of the model's hat matrix and is analogous to model degrees of freedom in linear regression. The model degree of dependency (*D*) is a metric that ranges from 0 to 1 and indicates the overall level of dependency in the local parameter estimates for the model. The formulas for *D* and D^* are given in Chapter 3.

Optimized covariate-specific bandwidths are shown under 'Bandwidths' in part 3 together with covariate-specific values of the effective number of parameters 'ENP_j', adjusted *t*-value at 95% confidence interval '*Adj t-val (95%)*', and covariate-specific degree of dependency 'D_j'. ENP_j is calculated based on the trace of the covariate-specific hat matrix and the values sum to the total effective number of parameters ($ENP = \sum_j^k ENP_j$). A corrected critical *t*-value for multiple hypothesis testing is also provided. Instead of using the nominal critical value of 1.96 (where $p = 0.05$), users can draw inference for each local parameter estimate using the adjusted *t*-value for each covariate. At the 0.05 level, any local estimate with an absolute value of local *t*-value less than the adjusted critical *t*-value is not statistically significant. The covariate-specific D_j, like the model *D*, quantifies the degree of dependence in the local parameter estimates, and its formula is provided in equation (3.9). Further details about these statistics are described in Chapter 3. The summary output also includes summary statistics of the local parameter estimates; mean, standard deviation, minimum, median, and maximum.

5.8.5 Local Model Results

The exported MGWR local CSV table (*_results.csv) contains columns shown in Table 5.2, and a screenshot of the file can be seen in Figure 5.10.

This table can then be imported into mapping or GIS software to visualize the spatial distributions of the local parameter estimates and other local statistics. Users can also create a choropleth map by joining the table with a shapefile using a common identifier. The example in Figure 5.11 shows how to use the open-source GIS software QGIS v3.16 to produce a choropleth map for the local parameter estimates of the covariate '*percentage of people with a bachelor's degree*' by joining the shapefile with the MGWR output based on the county ID. It can be seen that as the proportion of residents within a county with a bachelor's degree increases, the proportion of people voting Democrat increases significantly, *ceteris paribus*, in every county of the United States, although the impact of education on favoring the Democratic Party is greater in the West and the Northeast.[2]

TABLE 5.2

Columns in the Exported MGWR Local Statistics Table.

Column	Description	Column	Description
<ID>	ID field	localR2	Local R^2 values
x_coor	X coordinate of location	beta_<covariate>	Local parameter estimates for each covariate
y_coor	Y coordinate of location	se_<covariate>	Local standard errors for each covariate
y	Original dependent variable	t_<covariate>	Local *t*-values for each covariate
ols_residual	OLS residuals	p_<covariate>	Local *p*-values for each covariate
mgwr_yhat	MGWR prediction	sumW_<covariate>	Sum of local weights for each covariate
mgwr_residual	MGWR residuals		

MGWR_session_election_results.csv

county_id	x_coor	y_coor	y	ols_residual	mgwr_yhat	mgwr_residual	localR2	beta_intercept	beta_sex_ratio	...	sumW_pct_insured
1001	837137.896	-833746.195	-0.631	-0.891	-0.674	0.043	0.989	-0.693	0.008		13.682
1003	759093.379	-1051545.136	-0.911	-1.172	-0.819	-0.092	0.983	-0.643	0.000		13.403
1005	957282.206	-898765.288	0.799	-0.017	0.874	-0.075	0.976	-0.717	0.003		13.216
1007	788866.385	-784264.430	-0.813	-0.591	-0.769	-0.044	0.991	-0.644	0.009		13.721
1009	827160.381	-665520.377	-1.646	-0.965	-1.555	-0.091	0.980	-0.491	0.008		12.934
1011	925253.413	-875362.691	2.600	1.006	2.466	0.134	0.980	-0.702	0.005	...	12.967
1013	842709.914	-924527.214	0.546	-0.234	0.592	-0.046	0.986	-0.722	0.004		14.665
1015	894531.692	-683083.721	-0.371	-0.598	-0.266	-0.106	0.984	-0.498	0.010		14.116
1017	943719.963	-778213.843	0.499	-0.010	0.511	-0.012	0.983	-0.584	0.007		12.791
1019	908816.871	-634071.241	-1.258	-0.543	-1.285	0.027	0.977	-0.420	0.008		14.023
1021	826757.960	-798204.796	-1.166	-0.739	-1.101	-0.066	0.990	-0.662	0.009		14.220

FIGURE 5.10
A Snippet of the Exported MGWR Local Statistics Table.

5.9 Calibrating an MGWR Model Using the *MGWR* Python Library

In addition to MGWR 2.2, users can also use the open-source python package *mgwr*. Here, we present an example on how to call MGWR functions in Python, and detailed documentation can be found at https://mgwr.readthedocs.io/.

The *mgwr* package can be installed from the Python packaging index (PyPI) using the pip package manager:

```
pip install mgwr
```

To obtain in-development features, the user can also install *mgwr* directly from the source Github repository:

```
pip install https://github.com/pysal/mgwr.git
```

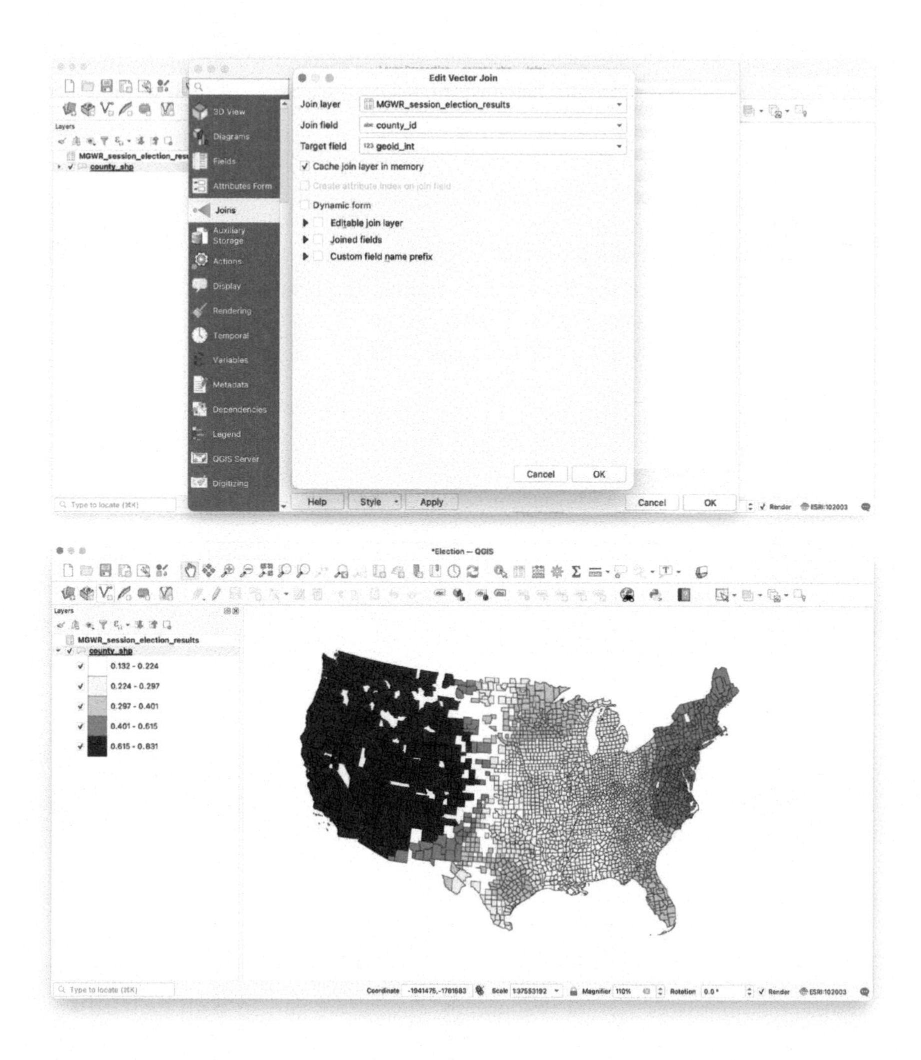

FIGURE 5.11
Joining and Visualizing Local Parameter Estimates in QGIS.

After installation, classes and functions can be imported from *mgwr* as well as from other common data handling packages such as *pandas*, *geopandas*, and *numpy*.

```
import pandas as pd
import geopandas as gpd
import numpy as np
from mgwr.gwr import GWR, MGWR
from mgwr.sel_bw import Sel_BW
```

After importing the necessary packages, data that conform to the data format specified in Section 3.2 are read. Here we provide an example of reading data from a CSV table with *pandas*.

```
data = pd.read_csv("us_election_2016.csv")
```

In a manner similar to that for MGWR 2.2 described earlier, users need to specify the model variables from the loaded data frame. Required variables include coordinates, the dependent variable, and independent variables, all of which must be converted to 2-D NumPy arrays before model fitting.

```
#Specify location variable
coords = np.array(list(zip(data[["proj_X"]].values,
                           data[["proj_X"]].values)))

#Specify dependent variable
y = data[["pct_dem"]].values

#Specify covariates
X_names = ['sex_ratio','pct_black', 'pct_hisp', 'pct_bach',
           'income', 'pct_65_over','pct_age_18_29', 'gini',
           'pct_manuf', 'log_pop_den','pct_3rd_party','turn_out',
           'pct_foreign', 'pct_insured']
X = data[X_names].values
```

MGWR requires standardization of both dependent and independent variables. To achieve this, the mean is subtracted and the result is divided by the standard deviation using NumPy functions.

```
#Variable Standardization
y = (y - y.mean())/y.std()
X = (X - X.mean(axis=0))/X.std(axis=0)
```

With the completion of data preparation now, users can fit both the GWR and MGWR models using similar function calls. There are two steps involved: (i) find the optimal bandwidth and (ii) use the optimized bandwidth to fit the model. In the case of GWR, the example below explains how to search for a single bandwidth and then fit the GWR model. The functions allow for arguments to modify kernel and search options. In this example, an adaptive bi-square kernel is used with a golden-section bandwidth optimization routine. For other kernel and search options, please refer to the full documentation.

```
gwr_selector = Sel_BW(coords, y, X, kernel="bisquare", fixed=False)
gwr_opt_bw = gwr_selector.search(search_method="golden_section",
                criterion="AICc")
print("GWR Optimal Bandwidth Found:", gwr_opt_bw)
```

```
GWR Optimal Bandwidth Found: 156.0
```

The returned optimal bandwidth gwr_opt_bw, which is 156 nearest neighbors, can then be used to fit a GWR model. The kernel settings used for the GWR fitting must be the same as those used for the bandwidth selection.

```
gwr_rslt = GWR(coords, y, X, bw=gwr_opt_bw, kernel="bisquare",
                  fixed=False).fit()
```

The syntax for bandwidth optimization in an MGWR model is identical to that in a GWR model, with the exception of one additional parameter: the user must include multi=True in the bandwidth selector Sel_BW().

```
mgwr_selector = Sel_BW(coords, y, X, kernel="bisquare", fixed=False,
                          multi=True)
opt_bws = mgwr_selector.search(search_method="golden_section",
                 criterion="AICc")
print("MGWR Optimal Bandwidths Found:", opt_bws)
```

```
MGWR Optimal Bandwidths Found: [43. 603. 43. 543. 208. 2659. 656. 58. 763.
2809. 387. 160. 117. 1424. 43.]
```

After a successful execution, a list of covariate-specific optimal bandwidths is returned, and the user must include the entire bandwidth selector object mgwr_selector (instead of a list of bandwidths) to initialize an MGWR object for fitting.

```
mgwr_rslt = MGWR(coords, y, X, selector=mgwr_selector, kernel="bisquare",
                  fixed=False).fit()
```

The fitted results for GWR and MGWR are saved in gwr_rslt and mgwr_rslt variables, respectively, and the following Table 5.3 shows some common properties and methods that can be called for both models. The full list can be found in the documentation.

The summary function offers a convenient way to summarize important statistics and inferences for the model, with outputs same to the summary file generated by MGWR 2.2 software.

```
mgwr_rslt.summary()
```

To visualize the local parameter estimates map, users can join the mgwr_rslt fitted object to a GIS file (e.g.,. shp,. geojson) using the common ID column and then plot the map using geopandas and matplotlib. The following is an example code snippet for visualization by defining a plot function. A shapefile containing the boundaries of US counties is read and projected to the NAD 1983 Albers North America projection (EPSG code: ESRI 102008). To overlay a state-level boundary, the dissolve function can be used on the basis of state ID 'STATEFP' in the data.

TABLE 5.3
GWR/MGWR Result Object Properties, Methods, and Their Descriptions.

GWR/MGWR result object attributes and methods	Descriptions
mgwr_rslt.predy	Predicted dependent variable
mgwr_rslt.resid_response	Model residuals
mgwr_rslt.ENP	Model effective number of parameters
mgwr_rslt.R2	Model R^2
mgwr_rslt.aicc	Model *AICc*
mgwr_rslt.params	Local parameter estimates
mgwr_rslt.bse	Local standard errors of estimates
mgwr_rslt.tvalues	Local *t*-values
mgwr_rslt.localR2	Local R^2 values
mgwr_rslt.summary()	Print model summary
mgwr_rslt.spatial_variability(mgwr_selector)	Perform Monte Carlo spatial variability test
mgwr_rslt.critical_tval()	Compute critical *t*-value
mgwr_rslt.local_collinearity()	Compute local collinearity measures
mgwr_rslt.get_bws_intervals(mgwr_selector)	Compute bandwidths confidence intervals

```
#Read a shapefile and apply a map projection
shp = gpd.read_file(county_shp.shp")
shp = shp.to_crs('ESRI:102008')
#Dissolve counties to obtain state boundary
state = shp.dissolve(by='STATEFP').geometry.boundary
```

The geopandas library, together with the matplotlib and mgwr.utils mgwr utility functions, can be used to visualize the parameter estimate surface.

```
import matplotlib as mpl
from mgwr.utils import shift_colormap, truncate_colormap
from matplotlib import cm,colors
```

Next, a plotting function can be defined to visualize the parameter estimate surface for any given independent variable with column index *j*. Within the function, a preferred colormap must be specified. In this example, a red-to-white-to-blue colormap is used to reflect the political context. The same colormap then will be used to generate a colorbar as the legend to be placed to the right of the map. The shapefile GeoDataFrame data can then be plotted using the values from the MGWR results, specified by the column=mgwr_rslt.params[:,j] argument in the GeoDataFrame.plot() function. The user can also filter out any insignificant parameters as gray color if they exist for some independent variables.

```
def plot_local_est(j):

    #Specify legend range and color vmin = -1
    vmax = 1
    cmap = cm.get_cmap("bwr_r")
    cmap = shift_colormap(cmap, start=0.0,
                          midpoint=1 - vmax/(vmax + abs(vmin)), stop=1.)

    #Plot local parameter estimates
    ax = data.plot(column=mgwr_rslt.params[:,j],vmin=vmin, vmax=vmax,
                   figsize=(10,10),cmap=cmap,linewidth=0.1,
                   edgecolor='white')

    #Filter out any insignificant estimates
    mgwr_filtered_t = mgwr_rslt.filter_tvals(alpha=0.05)
    if (mgwr_filtered_t[:,j] == 0).any():
        data[mgwr_filtered_t[:,j] == 0].plot(color='lightgrey', ax=ax,
                                             linewidth=0.1,edgecolor='white')

    #Overlay state boundary
    state.plot(color=None,ax=ax,linewidth=0.5,edgecolor='black')

    plt.axis('off')

    #Add the legend
    fig = ax.get_figure()
    cax = fig.add_axes([0.99, 0.2, 0.025, 0.6])
    sm = plt.cm.ScalarMappable(cmap=cmap,
                               norm=colors.Normalize(vmin=vmin, vmax=vmax))
    sm._A = []
    fig.colorbar(sm, cax=cax)
```

The plotting function plot_local_est() defined earlier can then be called, for example, to visualize the parameter estimate surface for the percentage of people with a bachelor's degree (as shown in Figure 5.12), which has a column index of 4 in the data.

```
plot_local_est(4)
```

5.10 Other Software for Geographically Weighted Regression Modeling

It is worth noting that GWR modeling is also accessible on other platforms; however, they are limited in functionality and computational efficiency compared to MGWR 2.2. For instance, GWR calibration is available in GWR 4 (Nakaya et al., 2009), ArcGIS Pro 3.0,[3] *GWmodel* R library (Gollini et al., 2015), and *spgwr* R library

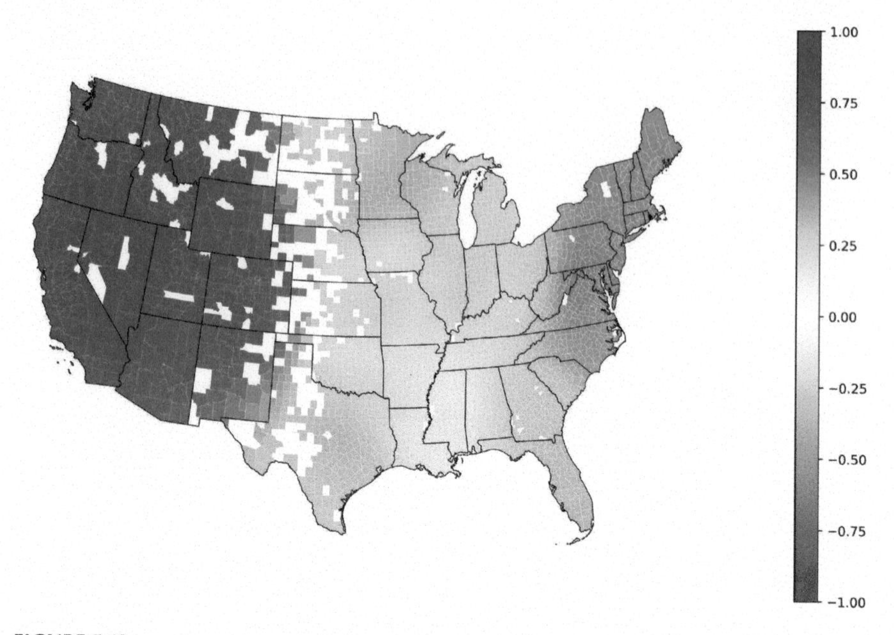

FIGURE 5.12
An Example Map of Local Parameter Estimates Associated With the Covariate "*percentage of people with a bachelor's degree*" Produced by Python.

(Bivand et al., 2013). Of these, *GWmodel* and ArcGIS Pro allow the calibration of MGWR models, both of which offer basic MGWR functionality that implements the estimation routine and inference computation from Fotheringham et al. (2017), Oshan et al. (2019), and Yu et al. (2020). However, at the time of writing, they lack recent methodological advancements such as bandwidth confidence intervals (Li et al., 2020), the computation of dependency measures for correcting significance tests, and computational improvement (Li et al., 2019; Li & Fotheringham, 2020). MGWR 2.2, along with the *mgwr* python package, explicitly focuses on the multiscale analysis of spatially heterogeneous processes, and we anticipate that this will be a long-term open-source project, with new developments being integrated into the software as they become available in the literature. Future software development and feature enhancement of MGWR 2.2 will be available at *https://sgsup.asu.edu/sparc/mgwr*.

5.11 Summary

In this chapter, we demonstrate the computational details and the usage of MGWR 2.2 software to calibrate an MGWR model. We strongly recommend that users of the software not only cite the name and location of the software but also explicitly report the software settings used in any calibration so that other researchers can reproduce the results if needed. The settings that should be reported are (i) the name and version of the software used; (ii) whether the data were standardized; (iii) the type of spatial kernel used; (iv) the choice of kernel function; (v) the bandwidth search method; and (vi) the optimization criterion employed.

Notes

1. The *local VIF* statistic is not available for MGWR currently because the calculation involves a single W matrix whereas MGWR computes covariate-specific W matrices. The statistic is available in GWR calibrations.
2. The unshaded counties in Figure 5.11 are those with fewer than 5,000 inhabitants and which were excluded from the analysis on the basis that their percentages might be unstable.
3. https://pro.arcgis.com/en/pro-app/latest/tool-reference/spatial-statistics/how-multiscale-geographically-weighted-regression-mgwr-works.htm

References

Belsley, D. A., Kuh, E., & Welsch, R. E. (2005). *Regression diagnostics: Identifying influential data and sources of collinearity.* New York: Wiley.

Bivand, R., Yu, D., Nakaya, T., Garcia-Lopez, M. A., & Bivand, M. R. (2013). Cran package 'spgwr'. *R Software Package*, 1–21.

Cressie, N. (1993). *Statistics for spatial data.* New York: Wiley.

da Silva, A., & Fotheringham, A. S. (2016). The multiple testing issue in geographically weighted regression. *Geographical Analysis*, *48*(3), 233–247.

Fotheringham, A. S., Li, Z., & Wolf. L. J. (2021). Scale, context, and heterogeneity: A spatial analytical perspective on the 2016 US Presidential election. *Annals of the American Association of Geographers*, *111*(6), 1602–1621.

Fotheringham, A. S., & Oshan, T. M. (2016). Geographically weighted regression and multicollinearity: Dispelling the myth. *Journal of Geographical Systems*, *18*(4), 303–329.

Fotheringham, A. S., Yang, W., & Kang, W. (2017). Multiscale geographically weighted regression (MGWR). *Annals of the American Association of Geographers*, *107*(6), 1247–1265.

Gollini, I., Lu, B., Charlton, M. E., Brunsdon, C., & Harris, P. (2015) GWmodel: An R package for exploring spatial heterogeneity using geographically weighted models. *Journal of Statistical Software*, *63*, 1–50.

Li, Z., & Fotheringham, A. S. (2020). Computational improvements to multi-scale geographically weighted regression. *International Journal of Geographical Information Science*, *34*(7), 1378–1397.

Li, Z., Fotheringham, A. S., Li, W., & Oshan, T. M. (2019). Fast geographically weighted regression (fastgwr): A scalable algorithm to investigate spatial process heterogeneity in millions of observations. *International Journal of Geographical Information Science*, *33*(1), 155–175.

Li, Z., Fotheringham, A. S., Oshan, T. M., & Wolf, L. J. (2020). Measuring bandwidth uncertainty in multiscale geographically weighted regression using Akaike weights. *Annals of the American Association of Geographers*, *110*(5), 1500–1520.

Nakaya, T., Fotheringham, A. S., Charlton, M. E., & Brunsdon, C. (2009). Semiparametric geographically weighted generalised linear modelling in GWR 4.0. *Proceedings of Geocomputation*, 1–5.

Oshan, T. M., Li, Z., Kang, W., Wolf, L. J., & Fotheringham, A. S. (2019). mgwr: A python implementation of multiscale geographically weighted regression for investigating process spatial heterogeneity and scale. *ISPRS International Journal of Geo-Information, 8(6)*, 269.

Yu, H., Fotheringham, A. S., Li, Z., Oshan, T. M., Kang, W., & Wolf, L. J. (2020). Inference in multiscale geographically weighted regression. *Geographical Analysis*, *52*(1), 87–106.

6

Caveat Emptor!

6.1 Introduction

To this point we have argued for the use of a local modeling framework, particularly that of MGWR although the arguments apply equally to other frameworks. The case has been made that *in certain circumstances* the processes being modeled might exhibit spatial nonstationarity and then a global model would be inappropriate because (i) it would lead to the incorrect inference that the processes are stationary over space; (ii) it is highly likely that inferences about the parameter estimates would be incorrect because the residuals from the model are likely to be spatially autocorrelated; (iii) the parameter estimates may contain a misspecification bias because contextual effects are not accounted for; and (iv) the model is likely to generate relatively poor replications or predictions of observed values. In this chapter we do not argue against the use of local models, but we make the case that one needs to be careful before applying them. The availability of user-friendly software for GWR, and now MGWR, as demonstrated in Chapter 5, has made it easy to apply these types of local models. However, a downside to this is that there have been many empirical applications that appear to lack credibility or which contain inferences that are not merited (Fotheringham, 2023). In order to reduce inappropriate usage of local modeling frameworks, we now describe a series of checks one should carry out prior to either undertaking the calibration of a local model or drawing conclusions from the results of such a calibration.

There are three sets of concerns anyone undertaking an empirical application using local models should address: (i) pre-calibration caveats; (ii) calibration caveats; and (iii) post-calibration caveats. The first set concerns the application itself in terms of what processes are being modeled and at what locations data are available to allow such modeling. The second set concerns the form of the model being calibrated, and the third set concerns the interpretations of the results. We now consider each of three sets of caveats.

DOI: 10.1201/9781003435464-6

6.2 Pre-Calibration Caveats

6.2.1 Does Local Modeling Make Sense?

As discussed in Chapter 1, the first question to ask before undertaking any type of local model calibration is "*Does this make sense?*" That is, is there any reason to suspect that the processes being modeled vary across space? In Chapter 1 we argued that the answer to this question could be "*yes*" in situations where context is thought to play a role in affecting either the local intercept or the way regressor attributes (covariates in a model) affect the dependent variable. We defined the former as "*intrinsic contextual effects*" and the latter as "*behavioral contextual effects*". Behavioral contextual effects are generally what we think of as creating spatially varying processes—there is some facet of location that makes the processes being modeled, and hence the estimated parameters in a model, vary over space. Intrinsic contextual effects largely capture the unmeasurable or non-measured variables that would be included in the model were it possible to do so. The omission of such variables is captured by a spatially varying local intercept.

This leads to three scenarios concerning the use of local modeling, as summarized in Table 6.1. The first is in physical systems controlled by laws where neither intrinsic nor behavioral contextual effects would apply. For instance, we do not attempt to estimate locally varying parameters in the Ideal Gas Law, Ohm's Law, or Newton's Law of Gravitation applied to inert bodies.[1] In such circumstances, local models have no role to play. The second is in human systems (those in which human values, preferences, and actions play important roles) where both intrinsic and behavioral contextual effects are plausible and where the use of local models should be considered. The third is in environmental systems, which perhaps provide the most controversial set of circumstances in which to consider using local models. On the one hand, it could be argued that there is little or no reason to think that behavioral contextual effects might be present in environmental systems—the determinants of ocean salinity, pine beetle density, or lake acidification would not seem to be unduly influenced by any type of behavioral contextual effects. However, such systems are difficult to model accurately, and there may well be intrinsic contextual effects caused by model misspecification that could be captured via a locally varying intercept.

TABLE 6.1

The Justification for the Use of Local Models in Different Systems.

	Intrinsic contextual effects (spatially varying intercepts)	Behavioral contextual effects (spatially varying slope parameters)
Human	✓	✓
Environmental	✓	✗
Physical	✗	✗

Consequently, users of local models should think very carefully about whether the type of application they are undertaking makes sense and what would account for any significant spatial variations in local parameter estimates. For instance, in environmental applications, it might be justifiable to estimate locally varying intercepts but not locally varying slope parameters; in human systems, it might be justifiable to estimate both; and in physical systems it is not justifiable to estimate either.

6.2.2 Data Issues

A separate set of questions about whether the application of a local model makes sense concerns the number of locations for which data are available and their spatial distribution. The calibration of a local model will only be useful if the locations for which data are recorded are both numerous and contain no major spatial discontinuities. However, both these criteria are inherently fuzzy, and there is no clear-cut definition of what constitutes a sufficient number of observations nor of what constitutes a major spatial discontinuity. What can be said is that the inferences drawn from calibrating a local model with data from a small number of locations some of which are spatially distant from the rest will be much less robust than those drawn from a large number of locations, which are spatially compact.

In terms of the number of locations, we recommend that there be at least 500 in order to calibrate a local model. The reason for this can be seen by reference to Figure 1.4, which describes the trade-off between uncertainty and bias in the local parameter estimation. For each local calibration, there needs to be a reasonable number of observations in order for the standard errors of the local parameter estimates to decrease more slowly than the bias increases as more observations are added to the local calibration. The point at which these gradients are equal denotes the optimal bandwidth. Recalling that the data used in each local calibration are weighted between 1 and 0, there needs to be at least 40 observations in each local calibration for the parameter estimates to be reasonably robust. If there were say only 100 observations in total, each local model would be very similar to its neighbor, and there would be relatively little spatial variation in the local parameter estimates even if the processes the parameter estimates represent varied greatly. Consequently, in order to be able to measure spatial process nonstationarity accurately, the sample set of locations for which data are available should be as large as possible and at least 500. The calibration of a local model is possible with fewer than 500 observations, but the resulting maps of the local parameter estimates will exhibit only broad regional trends at best.[2] In summary, we can state the following guideline: *For any given area of interest, the greater the density of locations for which data are available, the more local variation in processes can be identified should such variation exist.*

In terms of the spatial distribution of locations, it is preferable to have a set of locations, which do not have any obvious spatial discontinuities. For example, consider the two distributions of the same number of locations (500) in Figure 6.1. In Figure 6.1 (a) the distribution contains no obvious discontinuities (*good*), whereas in Figure 6.1 (b), there are obvious discontinuities between the separate clusters of locations (*bad*). The problem in the latter case is that if a cluster contains fewer

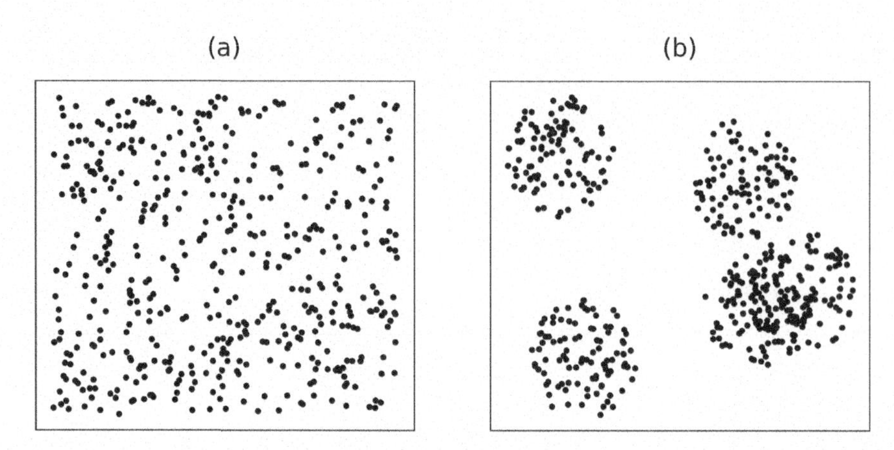

FIGURE 6.1
Two Sets of Locations: (a) Contiguous and (b) Noncontiguous.

locations than the optimal bandwidth (defined in terms of number of nearest neighbors), data for local calibrations would need to be borrowed from locations much further away. Given the relative distance between locations in the various clusters, these will tend to have very small weights creating greater uncertainty in the local parameter estimates. The classic example of this problem is when a local model is calibrated on US data, which contains Hawaii and Alaska. The solution is generally to remove the spatially discontinuous data from the analysis. Of course, if the number of locations in the smaller clusters is sufficiently large (i.e., larger than the optimal bandwidth), the model can still be calibrated and the results will be equivalent to running separate local model calibrations for each cluster.

A final caveat regarding the data used in any MGWR application is that both the dependent and independent variables should be standardized $N(0,1)$ prior to calibration. In the MGWR software described in Chapter 5, this is one of the default settings and has to be actively turned off. Without data standardization, the optimized covariate-specific bandwidths will be, in part, a function of the variability of each covariate, and to ensure comparability across bandwidths, it is therefore important to make the variability of each covariate the same. Data standardization also has the advantage that the magnitudes of the local parameter estimates can be compared directly across covariates. The resulting local parameter estimates and estimates of y_i can easily be converted to their unstandardized forms if this is required.

6.3 Calibration Caveats

In our second set of warnings we turn attention away from the data employed in the calibration of a local model and to the form of the model being calibrated.

6.3.1 The Global Model

The starting point for any successful local model calibration must be a good, robust, defensible, global model. Too many applications of local models have been published starting from a weak global model where "impressive" increases in goodness-of-fit are almost certainly due to the local model overfitting and the local intercept essentially becoming a proxy for the dependent variable. It is therefore imperative that the global model contain all the relevant explanatory variables that theory, common sense, and other empirical applications suggest are important. It is also imperative that various diagnostic checks are undertaken on the global model to ensure the multitude of potential problems in regression analysis—'the usual suspects'—are avoided or minimized. There are, however, two problems with these strictures.

The first is that what constitutes a "*good, robust, defensible*" global model is likely to be application-specific so that no general guidelines can be set. In some applications, such as in modeling the occurrence of certain types of disease, random factors can play a significant role so that global model performance appears weak despite models incorporating all of the obvious covariates. In other applications, random events are less influential, and global model performance is generally strong. Consequently, there can be no rule of thumb as to what level of R^2, say, should be reached by a global model prior to a local calibration: the decision must be based on common sense, and, relating to the discussion in Chapter 1, the general rule is to ensure that all *measurable* covariates are included in the model.

The second relates to how diagnostics checks are undertaken—should these be on the global model or the local model? The answer is that diagnostic checks should be undertaken on *both* the global and local models for different reasons. Various checks for issues such as correct model form and multicollinearity should be carried out on the global model prior to any local calibration but similar checks (see Sections 6.3.2 and 6.3.3 for more details) should also be undertaken on the local models because problems with the local calibrations may not be detectable in the global calibration. We now examine two potential problems of local modeling that relate to model form; one of which, nonlinearity, is particularly relevant to local models but has largely gone unheeded in the literature (nonlinearity), and the other, multicollinearity, although featuring fairly prominently in the literature on local models, is likely to be less of an issue when appropriate diagnostics are employed.

6.3.2 Nonlinearity Versus Spatial Process Heterogeneity

One of the main outputs from the calibration of a local model is a set of geocoded local parameter estimates that can be mapped to display any spatial heterogeneity in their values; sufficient heterogeneity can then be used to infer process spatial nonstationarity. However, as Sachdeva et al. (2022) demonstrate, suppose a linear model is calibrated locally, but the parameter estimates vary in covariate space (i.e., the relationships are nonlinear) and the covariate is distributed over space with some degree of spatial dependency. This would also result in spatially varying local parameter estimates even when the processes being modeled are stationary. For example, if the associated local parameter estimates are positive when covariate x is large and negative when x is small (representing a U-shaped nonlinear relationship between y and x) and x is distributed across space such that values are larger in the north than in the south, the resulting local parameters would have a distinctive spatial pattern being positive in the north and negative in the south. Consequently, it could be argued that spatial variations in parameter estimates reported in the calibration of local models might not be indicative of intrinsically spatially varying relationships, as claimed, but merely reflect nonlinear conditioned relationships between y and a particular covariate.

Sachdeva et al. (2022) report that this situation can be both easily identified and remedied. The obvious way to check the provenance of spatially varying local parameter estimates is to plot the local estimates, $\widehat{\beta}_{ji}$, against the respective value of the covariate, x_{ji}. If nonlinearity between a covariate x and the response variable y is the source of the spatially varying parameter estimates, there would be a clear structure to this plot. Conversely, if the resulting plot exhibited no discernable structure, it could be assumed that variation in the local estimates is attributable to a spatially nonstationary process. As a follow-up to any suspected nonlinear influence on locally varying parameter estimates, a vector from the estimated parameter values equal to $(\beta_{ji} * x_{ji})$ for each covariate, j, can be plotted on the y-axis against the covariate x_{ji} on the x-axis.[3] Again, if there is nonlinearity in the conditional relationship between y and x, some structure would be evident in this plot, which would suggest the nature of the nonlinearity to be modeled in a reformation of the local model.[4]

In order to examine the efficacy of the diagnostic test described earlier, Sachdeva et al. (2022) explore the calibration of both a local model, MGWR, and a model form designed to capture nonlinear relationships, a generalized additive model (GAM), in two experiments. In experiment 1 the processes being examined are nonstationary and the dependent variable is related to the independent variables through a set of linear relationships. In this experiment the diagnostic plot should indicate no discernible structure so that significant spatial variation in the local parameter estimates can be taken as an indication of spatially varying processes. In experiment 2 the processes being examined are stationary over space, but the independent variables have some predefined spatial dependency, and the dependent variable is related to the independent variables nonlinearly. In this experiment the diagnostic plot should reveal clear structure indicating that spatial heterogeneity in the local parameter estimates is caused by model misspecification, which should be corrected.

To demonstrate the application of the diagnostic plot, Sachdeva et al. construct a dataset with 2,500 locations arranged on a 50 × 50 grid and then calibrate a model of the following form:

$$y_i = \beta_{2i} x_{2i} + \beta_{2i} x_{2i} + \epsilon_i \tag{6.1}$$

Local surfaces for β_1, β_2, x_1, and x_2 were simulated using Gaussian random fields to control the degree of spatial dependency in the surfaces. Further details can be found in Sachdeva et al. (2022). In experiment 1, the dependent variable y was generated using equation (6.1) where both x_1 and β_1 and x_2 and β_2 were independent of each other representing the situation where there is no dependency between the parameter values and the covariate levels so that any observed spatial variation in the parameter estimates can be attributed solely to spatially varying processes. The diagnostic plots for the two sets of local parameters are shown in Figure 6.2 where it is obvious that there is no structure in either set of estimates. Any spatial variation in either set of estimates can then be attributed to a spatially nonstationary process.

In experiment 2, using the same 2,500 location points as earlier, the independent variables x_1 and x_2 are simulated using a Gaussian random field, but the parameter estimates are deliberately made a function of the respective covariate levels, which means the relationships being modeled are nonlinear and the local parameter estimates will reflect spatially varying values of the two covariates rather than spatially varying processes. The diagnostic plots appear in Figure 6.3 and clearly suggest that there is structure to both relationships so that the spatial variation in the local parameter estimates is a product of inappropriately modeling nonlinear relationships with a linear model. In this case, the spatial variation in the local estimates should not be interpreted as indicative of a spatial process nonstationarity; instead the model should be reformulated in a nonlinear form and recalibrated. A further example with greater detail of the deployment of this simple diagnostic check is given by Sachdeva et al. (2022) for a local modeling analysis of US Presidential voting.

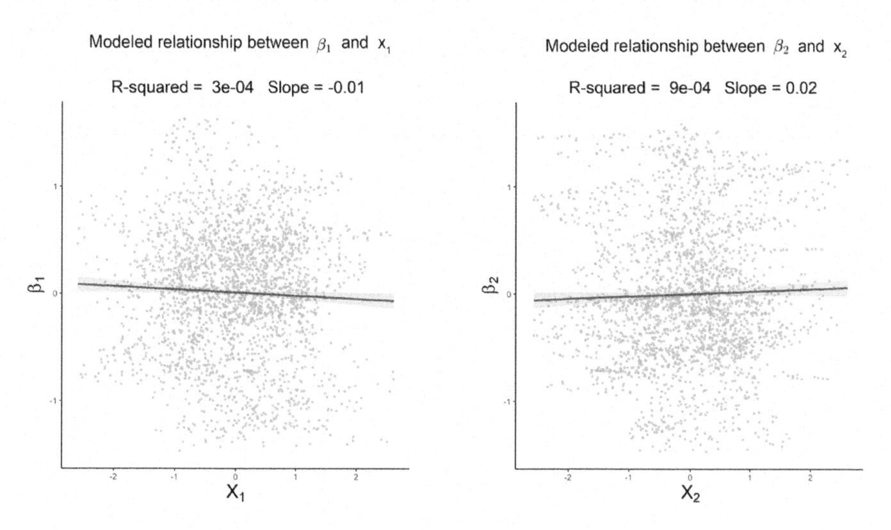

FIGURE 6.2
An Example of the Diagnostic Plots to Test for Nonlinearity as a Cause of Spatially Varying Parameter Estimates. (*The lack of structure indicates no evidence of any nonlinearity in the relationships so that the spatial variation in the parameter estimates can be interpreted in terms of spatial process nonstationarity.*)

Source: Sachdeva et al. (2022).

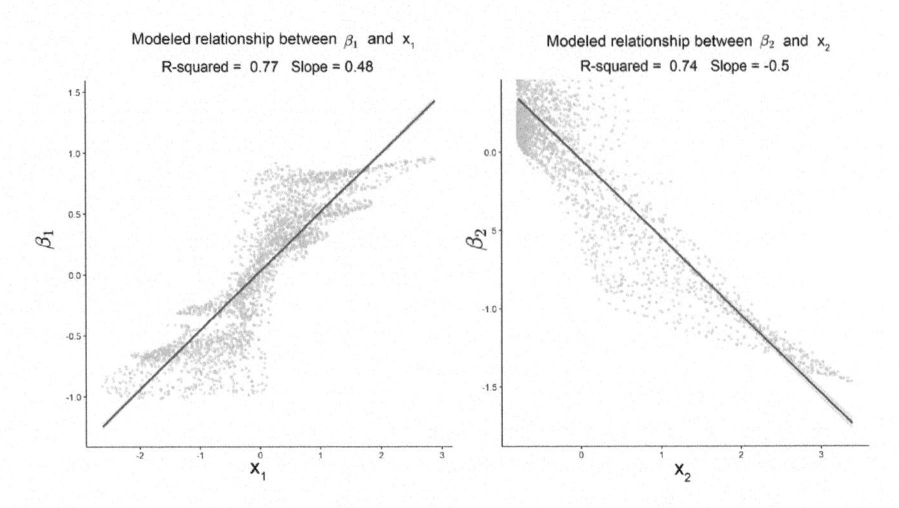

FIGURE 6.3
An Example of the Diagnostic Plots to Test for Nonlinearity. (*The clear structure indicates nonlinear relationships so the linear model is misspecified and the spatial variation in the parameter estimates cannot be interpreted in terms of spatial process nonstationarity.*)

Source: Sachdeva et al. (2022).

In summary, the diagnostic test shown here, which involves plotting the local parameter estimates against the level of the respective covariate, should form an essential part of any local modeling application. Without checking for nonlinearity, there is a risk of incorrectly attributing locally varying parameter estimates, which result from applying a linear form of model to nonlinear relationships, to spatially varying processes.

6.3.3 Multicollinearity in Local Models

Multicollinearity occurs in a model whenever two or more of the independent variables are *highly* correlated. The effect of very high correlation between two, or among several, independent variables is to cause confusion of the parameter estimates and inflate their standard errors (Belsey et al., 1980). In extreme cases, if two covariates essentially measure the same thing, the parameter estimate for one covariate will be large and positive while the other will be large and negative—both with huge standard errors. For details of parameter estimate sensitivity to multicollinearity in a linear model, see Fotheringham (1982). However, the problem is that there is no general agreement on how to define '*highly* correlated'—that is, how strongly do two or more variables have to be related in order for us to be suspicious of the resulting parameter estimates? Suggested diagnostics include simple correlation structures between variables or more complex measures such as condition numbers (*CN*), variance decomposition proportions (*VDP*), and variance inflation factors (*VIF*), but there is no general agreement on the critical values of any of these diagnostic statistics so that the decision on whether there is a significant problem of multicollinearity is largely subjective (*inter alia*, O'Brien, 2007).

A potential problem in local modeling, first identified by Wheeler and Tiefelsdorf (2005), is that the effects of multicollinearity might cause the local parameter estimates derived from GWR (MGWR was not developed at the time of their analysis) to exhibit spatial heterogeneity if different levels of multicollinearity are recorded between the same variables across the various local datasets used to calibrate the model. Although the multicollinearity "problem" in GWR has received some attention, and is often cited, it emanates from a flawed analysis. For reasons comprehensively described by Fotheringham and Oshan (2016), it is not an inherent issue, and the authors demonstrate that multicollinearity only becomes an issue in GWR when higher levels of similarity between two or more variables are encountered. Furthermore, undertaking the appropriate local diagnostics can help avoid any issues that might arise. Multicollinearity is typically less of a problem in MGWR because the covariate-specific bandwidths mean that different covariates will have different weights for the same locations and will contain measurements taken from different sets of locations.

Nonetheless, various multicollinearity diagnostics have been developed for local regression models such as local variants of variance inflation factors, condition indices, and variance decomposition proportions (see Chapter 2 for more details and Chapter 5 for examples of these in MGWR 2.2). In addition, Wheeler (2010) presents visualization tools, such as linked scatter plots and parallel coordinate plots, for determining potential areas where local multicollinearity may be problematic. These tools seek to inform the analyst where multicollinearity may be an issue for any individual local models within GWR, with the goal of ultimately treating them through the use of additional remedial methods. Two extensions to the GWR framework that have been proposed are ridge regression (Wheeler, 2007; Barcena et al., 2014) and LASSO regression (Wheeler, 2009). Both of these methods constrain parameter estimates based on their variance. However, while these techniques may result in slightly lower prediction errors in some instances, they entail sacrificing model fit as well as requiring more complex algorithms that result in significantly increased computation time. Given that multicollinearity problems in MGWR are likely to be quite rare and should be easily spotted given the local multicollinearity diagnostics available and corrected at the global modeling stage, the equivalent algorithms for MGWR do not exist.

Local sample augmentation is another proposed extension to GWR to offset any issues with local multicollinearity. Brunsdon et al. (2012) suggest increasing the bandwidth for a given location in a GWR model where high local multicollinearity is indicated. Barcena et al. (2014) take this idea a step further by developing a framework for enriching each local sample in a manner that optimizes the local design matrix with additional observations from the larger study area. They demonstrate the possibility of selecting enrichment observations from the entire study area, as well as restricting the selection of enrichment observations to relatively nearby locations in order to decrease local *VIF* values. While the method is shown to be successful in counteracting multicollinearity without greatly modifying existing calibration routines, there are some potential drawbacks. Certain observations may become highly influential if they are repeatedly selected in the enrichment of many locations. Additionally, if the selected enrichment observations are relatively far from the calibration location, it would prove difficult to provide a sound theoretical basis for the inclusion of these observations.

An issue that persists throughout much of the literature on GWR and multicollinearity is that the examples used, whether with real or simulated data, include only a relatively small sample size within the study region (Paez et al., 2011). Using only a small number of spatial units in local modeling is bad practice (see Section 6.2.2) for other reasons, but it also makes the local calibration more susceptible to multicollinearity since there are relatively few neighbors contributing to repeated weighted observations for each location. To counter this issue, Paez et al. (2011) carry out simulation experiments using varying sample sizes in order to investigate GWR's ability to replicate known parameter surfaces under varying degrees of multicollinearity. They report that GWR provides relatively accurate parameter estimates for stationary and nonstationary process subject to different levels of multicollinearity among explanatory variables, especially for larger sample sizes. These results are supported in Fotheringham and Oshan (2016) using a more comprehensive set of simulations.

Another theme, conspicuously missing from the earlier literature on GWR and multicollinearity, relates to the concept of significance testing (see Chapter 3). It is entirely possible that any counterintuitive trends in local parameter estimates reported in earlier studies actually pertained to insignificant parameter estimates that should not have been meaningfully interpreted. As mentioned, one of the effects of multicollinearity is to increase the standard errors of the local parameter estimates. The need for appropriate significance testing procedures to be undertaken on any local modeling results is stressed later, and the appropriate techniques to do this are described in detail in Chapter 3.

In summary, there might be good reasons for not undertaking the calibration of a model by MGWR, but as long as the appropriate checks are conducted, multicollinearity should rarely, if ever, be one of them.

6.4 Post-Calibration Caveats

6.4.1 Inference

As discussed in Chapters 1 and 3, inference is hugely important in local modeling as it is very easy to develop plausible-sounding explanations for patterns of local parameter estimates, which are essentially a product of noise. Assuming that the parameter estimates obtained from a local model are free from serious misspecification biases, spatial variations in the estimates can result from either, or both, sampling variation and intrinsically different processes being modeled. It is therefore essential to check that the variation in the local estimates is sufficiently great that the former can be discounted so that the estimates may be interpreted as indicating spatially varying processes. This can be done by applying the various tests described in Chapter 3. Here, we merely reiterate that there are three types of inferential tests that need to be applied to the results of calibrating a model by MGWR:

1. Inference about the *individual* local parameter estimates. Here it is important to account for both multiple hypothesis testing and the degree of dependency in the local estimates (see Chapter 3 for details).
2. Inference about the overall variability of the *set* of local estimates. Even when the processes being modeled are stationary over space, local parameter estimates will still exhibit spatial variability. The important question that must be answered in any local modeling application is then: *Is the amount of spatial variability sufficiently large to allow the contribution of sampling variation to be discounted?* Only if this question has been answered affirmatively can we look for reasons why the processes being modeled might vary spatially.
3. Inference connected to the covariate-specific optimal bandwidths. As shown in Chapter 3, the optimal bandwidths produced in the calibration of MGWR models have uncertainty attached to them which can be quantified and a confidence interval placed around each covariate-specific bandwidth. This can be used to compare the spatial scales over which different sets of local parameter estimates vary and hence the different levels of spatial nonstationarity exhibited by the processes the parameters represent.

A final issue regarding the reporting of the results of a local model calibration is what to do when a global parameter estimate is statistically significant but the majority of the local estimates of that parameter are not, or when a global estimate is not statistically significant but the majority of the local estimates are. Fotheringham and Sachdeva (2022) demonstrate extreme cases of this with full reversals in signs between the global and local models—a spatial variant of the well-known Simpson's paradox. Of course, both results can be correct because the models are applied at different spatial scales and therefore answer different questions. For instance, the application of a global hedonic price model, which models house prices across a city, could yield evidence that newer houses generate a premium over older houses, *ceteris paribus*, whereas a local model applied to the same data could yield evidence that older houses are worth more than newer ones, *ceteris paribus*. In the global application the scale of the comparison is across the whole city where newer neighborhoods may be more attractive than older neighborhoods. In the local application, the comparison is within local neighborhoods where older houses might have a certain cachet and be more attractive than newer ones. This is a scale issue—in the former application neighborhoods within a city are being compared; in the latter, houses are being compared within neighborhoods. As with aspatial examples of Simpson's paradox, there is no right or wrong answer: the researcher has to recognize that because of the different spatial scale of the applications, different questions are being answered.

6.4.2 Visualization

The outputs from MGWR lend themselves to visualization, typically, but not exclusively, in terms of geocoded local parameter estimates and associated diagnostics. Two issues need to be addressed prior to the visualization of these outputs: (i) should the nonsignificant parameter estimates be mapped and (ii) what happens if two or more data points share the same location, and hence their local parameter estimates share the same coordinates? There is no absolute answer to the former, and it is largely up to personal choice. However, it is generally useful to distinguish between the significant and nonsignificant local estimates either by only plotting the significant values or by masking the nonsignificant ones or by plotting the associated *t*-values and color-coding these according to the level of significance used.[5] This prevents the researcher being tempted to develop narratives about trends in the nonsignificant local estimates, which may well be the product of noise. The latter problem can be overcome by jittering any shared coordinates.

6.5 Summary: An MGWR Model Calibration Checklist

No empirical application is perfect. Many decisions that have to be taken in the calibration of a model are subjective and hence can be the subject of criticism or debate. Hence, it is not the intention here to provide a list of actions that will lead to the perfect MGWR calibration. Rather, as a summary to the above discussion, we provide a set of seven questions, which, if answered affirmatively, will lead to a better empirical application.

1. Does the application of a local model make sense?
2. Do the data, in terms of both the number of observations and their spatial distribution, allow a sensible model calibration?
3. Is the global model robust in terms of the variables included and their functional form? Have appropriate checks been undertaken to ensure this?
4. Has a check been undertaken to ensure that nonlinearity is not a possible cause of spatially varying parameter estimates rather than spatially varying processes?
5. Have all the data been standardized N(0,1) prior to calibration?
6. Have suitable inferential tests been undertaken on both the individual local estimates and also the local parameter surfaces? The former should include corrections for multiple hypothesis testing and dependency.
7. If the local parameter estimates are depicted on a map, have nonsignificant estimates been suitably identified?

Notes

1. Interestingly, when Newton's Law of Gravitation (or any spatial interaction model based initially on an analogy to Newton's model) is applied to flows of people, goods, and information (all human-related activities), spatial variations in the distance-decay parameter are a noted feature of such applications (Fotheringham, 1981).
2. For demonstration purposes only, we occasionally make use of a dataset from Georgia on educational attainment rates by county, which contains only 159 observations.
3. These plots can easily be produced from the relevant columns in the data file (for the covariate levels) and the output file from MGWR 2.2 (for the local parameter estimates) although future developments of the MGWR software package (see Chapter 5) may include an option to produce such plots automatically.
4. The choice of nonlinear form is up to the user and usually involves a logarithmic or exponential form, but the plot of $(\beta_{ji} * x_{ji})$ against the covariate x_{ji} can be a useful indicator of the type of nonlinearity involved.
5. Of course, the level of significance used is arbitrary and hence so is the distinction between "significant" and "nonsignificant" parameter estimates, but we share the conviction that it is still useful to make some separation between those estimates, which provide relatively strong evidence of something other than noise and those that do not.

References

Barcena, M. J., Menandez, P., Palacios, M. B., & Tusell, F. (2014). Alleviating the effect of collinearity in geographically weighted regression. *Journal of Geographical Systems*, *16*(4), 441–466.

Belsey, D. A., Kuh, E., & Welsch, R. E. (1980). *Regression diagnostics: Identifying influential data and sources of collinearity*. New York: Wiley.

Brunsdon, C., Charlton, M. E., & Harris, P. (2012). *Living with collinearity in local regression models*. http://eprints.maynoothuniversity.ie/5755/

Fotheringham, A. S. (1981). Spatial structure and distance-decay parameters. *Annals of the Association of American Geographers*, *71*(3), 425–436.

Fotheringham, A. S. (1982). Multicollinearity and parameter estimates in a linear model. *Geographical Analysis*, *14*(1), 64–71.

Fotheringham, A. S. (2023). A comment on 'A route map for successful applications of geographically-weighted regression': The alternative expressway to defensible regression-based local modelling. *Geographical Analysis*, *55*(1), 191–197.

Fotheringham, A. S., & Oshan, T. M. (2016). Geographically weighted regression and multicollinearity: Dispelling the myth. *Journal of Geographical Systems*, *18*(4), 303–329.

Fotheringham, A. S., & Sachdeva, M. (2022). Scale and local modeling: New perspectives on the modifiable areal unit problem and Simpson's Paradox. *Journal of Geographical Systems*, *24*(3), 475–499.

O'Brien, R. M. (2007). A caution regarding rules of thumb for variance inflation factors. *Quality and Quantity*, *41*(5), 673–690.

Paez, A., Farber, S., & Wheeler, D. C. (2011). A simulation-based study of geographically weighted regression as a method for investigating spatially varying relationships. *Environment and Planning A*, *43*(12), 2992–3010.

Sachdeva, M., Fotheringham, A. S., Li, Z., & Yu, H. (2022). Are we measuring spatial non-stationarity or nonlinearity. *Geographical Analysis*, *54*(4), 715–738.

Wheeler, D. C. (2007). Diagnostic tools and a remedial method for collinearity in geographically weighted regression. *Environment and Planning A*, *39*(10), 2464–2481.

Wheeler, D. C. (2009). Simultaneous coefficient penalization and model selection in geographically weighted regression: The geographically weighted lasso. *Environment and Planning A*, *41*(3), 722–742.

Wheeler, D. C. (2010). Visualizing and diagnosing coefficients from geographically weighted regression models. In B. Jiang & X. Yao (Eds.), *Geospatial analysis and modelling of urban structure and dynamics* (pp. 415–436). Berlin: Springer.

Wheeler, D. C., & Tiefelsdorf, M. (2005). Multicollinearity and correlation among local regression coefficients in geographically weighted regression. *Journal of Geographical Systems*, *7*, 161–187.

7

A Local Analysis of Voting Behavior

The 2020 US Presidential Election

7.1 Introduction

As part summary of the previous chapters, and part showcase for how MGWR can be used to provide insights into contextual influences on human preferences and behavior, we now present an empirical example of the application of an MGWR model to a set of decisions by 159,633,396 people that led to a major domestic and international upheaval. In 2020, a presidential election was held in the United States, which saw the Republican Party, led by Donald Trump, ousted from government in favor of the Democratic Party, led by Joe Biden. This represented the largest voter turnout (66.7% of the eligible population voted) in 120 years of US Presidential election voting history. Of the votes cast, Biden received just over 81 million or 51.3%; Trump received just over 74 million, or 46.8%, and Third Party candidates (who varied from state to state) received almost three million or 1.8%. Despite the relatively wide margin of victory in terms of the popular vote, the influence of the Electoral College in US Presidential elections, whereby there is generally a "winner takes all" outcome on a state-by-state basis, meant that the outcome of the election was fairly close, a feature of several recent Presidential elections and one which is likely to continue into the foreseeable future. Consequently, it is important not only to understand what factors affect voters' preferences but also to determine if the responses in terms of voting behavior to each of these factors varied across the country. In particular, it is of interest and concern to determine what, if any, role geographic context plays in influencing how people vote. As explained in Chapter 1, there are two roles that context can play in determining human preferences and behavior: intrinsic contextual effects are the result of various unmeasurable influences on behavior being omitted from the model; behavioral contextual effects represent the role of location in influencing the way in which various socioeconomic factors affect behavior. Ideally, we want to separate the influence of both contextual effects and measure the influence of intrinsic contextual effects compared to the influence of socioeconomic factors in affecting behavior—in this case, how people vote in Presidential elections.

DOI: 10.1201/9781003435464-7

7.2 A Tale of Two Counties

In order to focus on the issue of context versus composition in terms of how people voted in the 2020 election, consider two particular counties in the United States: Iowa County, Wisconsin, and Panola County, Texas. The socioeconomic composition of both counties, described in Table 7.1, is very similar. Indeed, despite some differences discussed later, this is one of the pairs of counties across the whole of the United States that are most similar in terms of their socioeconomic composition and size.

Both counties have populations of just over 23,000 and very similar population densities. Both have almost exactly the same age profiles and similar levels of females to males. Both have largely white, stable populations, although Panola Co. has a larger proportion of African-Americans (17% vs. 1%) and a larger proportion of Hispanics (9% vs. 2%). Education levels and income levels are higher in Iowa Co. (25% have a bachelor's degree in Iowa Co. compared to 16% in Panola Co., and median household income in Iowa Co. is $64k compared to $53k in Panola Co.) There is a higher proportion of people without health insurance in Panola Co. (17% compared to 4%). Votes for a third-party candidate were about the same although turnout was higher in Iowa Co. than in Panola Co. (78% vs. 67%). Consequently, from simply observing the two profiles, one would guess that the voting behavior of the two counties was not hugely different, or one could be forgiven for thinking that the Democrat vote might be higher in Panola Co. given the higher proportions of Black and Hispanic voters, who tend to favor the Democratic Party, and also because

TABLE 7.1

Socioeconomic Composition of Iowa Co., WI, and Panola Co., TX.

Attribute	Iowa Co., Wisconsin	Panola Co., Texas
Population	23,618	23,327
Gender ratio	101	97
% Black	0.9	17.3
% Hispanic	1.8	8.7
% Bachelor's degree	24.6	16.1
Median income	64,124	52,982
% Ages 65 and over	18.0	18.7
% Ages between 18 and 29	12.0	14.2
Income disparity (gini)	0.4	0.5
% Manufacturing	13.3	6.0
Population density (log transformed)	3.4	3.4
% Third-party votes	1.8	0.6
Turnout	77.5	66.9
% Foreign-born	1.7	3.8
% With no health insurance	3.7	17.1

of the higher proportion of people without health insurance (typically, it would be in the self-interest of such people to vote Democrat). However, only 18% of Panola Co. voted Democrat, whereas 57% of Iowa Co. did. What then caused this massive difference in voting behavior?

The only explanations are that (i) some measurable attribute that has a huge influence on voting behavior is not included in Table 9.1, and, whatever this attribute might be, there must be a massive difference in the values of it, which strongly motivates people in Panola Co. to vote Republican and people in Iowa Co. to vote Democrat;[1] or (ii) there is something intangible about the two locations that promotes a different political leaning and which is independent of population composition. Here, we term this latter feature, *geographical context*, or simply *context* and, as mentioned earlier and in Chapter 1, this has two components: *intrinsic contextual effects* and *behavioral contextual effects*. Intrinsic contextual effects exist if where people live affects how they vote, and this place-based effect is independent of socio-demographics. Behavioral contextual effects exist if location affects the way individual characteristics determine how people vote. In this case, voters in Iowa Co. could be influenced by Democrat-leaning friends, work colleagues, and local media and possibly also by a historical link to the largely Scandinavian settlers of this area who left a legacy of social justice and placing the needs of the community before the individual. Voters in Panola Co., in eastern Texas close to the Louisiana border, on the other hand, could be influenced by Republican leaning friends, work colleagues, and local media and possibly also by a frontier-inspired mentality of independence whereby individuals are encouraged to be more self-sufficient and less reliant on central government. To explore these possibilities further, we now turn to a model of voting behavior for counties across the United States.

7.3 A Model of Voter Preference

Identifying the drivers of voter preferences for one party over another has clear and important implications not just for understanding the outcomes of electoral processes but also for being able to influence the outcomes of future elections by targeting key groups of voters. Consequently, a great amount of research has been undertaken to try to identify the factors that play a role in influencing how citizens vote (Sigelman & Sigelman, 1982; Powell, 1986; Hillygus & Jackman, 2003; Leigh, 2005; Mutz, 2018; Schaffner et al., 2018). Summarizing these and many other studies of voting behavior, and after a great deal of experimentation and testing, the 14 variables in Table 7.2 were identified as being important to include in any county-level model seeking to understand the division of voter preferences between the Democrat and Republican candidates in the 2020 US Presidential election. Further detail on each of these covariates and their justification for inclusion in the model can be found in Fotheringham et al. (2021) and Li and Fotheringham (2022).

TABLE 7.2

Covariates Used in a Model of Voter Preferences in the 2020 US Presidential Election.

Variable	Description
Sex_ratio	The ratio of males to females with 100 representing an equal number of males and females
Pct_age_18_29	Percentage of the population aged 18 to 29
Pct_age_65	Percentage of the population aged 65 and over
Pct_Black	Percentage of the population identifying as Black
Pct_Hispanic	Percentage of the population identifying as Hispanic
Median_income	Median household income
Pct_Bachelor	Percentage of the population with a bachelor's degree or higher
Gini	Gini index of income inequality; the greater the index value, the greater the income disparity
Pct_Manuf	Percentage of the population employed in the manufacturing sector
Ln(pop_den)	Natural logarithm of population density (persons per square mile). This acts as an urban/rural indicator
Pct_3rd_Party	Percentage of votes that went to a third-party candidate
Turnout	Voter turnout—the percentage of those eligible to vote who actually voted
Pct_FB	Percentage of the population identifying as being born outside the United States
Pct_Uninsured	Percentage of population without health insurance coverage

7.3.1 The Model

The dependent variable to be modeled is the percentage of the population in each county of the United States who voted for the Democratic Party in a straight contest between Democrats and Republicans. That is, any votes for Third Party candidates are ignored in the definition of the dependent variable, *%Dem*, which is

$$\%Dem = Dem * 100 / (Dem + Rep) \tag{7.1}$$

where *Dem* is the number of votes cast per county for the Democratic Party and *Rep* is the number of votes cast per county for the Republican Party. Two models are then calibrated with *%Dem* as the dependent variable and the 14 variables listed in Table 7.2 as the covariates. The first is a global model of the form:

$$y_i = \beta_0 + \beta_1 x_{1i} + \beta_2 x_{2i} + \ldots + \beta_k x_{ki} + \varepsilon_i \tag{7.2}$$

and the second is a local model of the form:

$$y_i = \beta_{0i} + \beta_{1i} x_{1i} + \beta_{2i} x_{2i} + \ldots + \beta_{ki} x_{ki} + \varepsilon_i \tag{7.3}$$

The former model is calibrated by OLS regression while the latter is calibrated by MGWR using adaptive covariate-specific bi-square spatial kernels whose bandwidths are optimized using *AICc* (see Chapter 2).

7.3.2 The Data

Large-scale voter preferences are typically recorded as the percentage share of the vote for one party within some aggregated spatial unit such as an electoral district or county (Kim et al., 2003; Levernier & Barilla, 2006; Scala et al., 2015). Such data will always have the disadvantage of being prone to the ecological fallacy whereby the results may not reflect the behavior of the individuals within the aggregated spatial unit. However, because the data tend to be much more comprehensive than, say, a survey of a limited number of individuals, they allow for more detailed spatial analysis. Given the focus here is on identifying the relative role of spatial context in determining voter preference for one party over another using spatially disaggregated models of voter choice, we employ county-level data on voter preferences in the 2020 US Presidential election.

The 2020 county-level election data were obtained from the MIT Election lab website.[2] The dependent variable in the regression models is defined as in equation (7.1) and is the percentage share of the vote that went to the Democratic Party out of the votes that went to either the Democratic Party or the Republican Party in each county. The spatial distribution of this variable for all US contiguous counties is shown in Figure 7.1 with counties in blue (red) being those where the majority of voters in the county voted for the Democratic (Republican) Party.

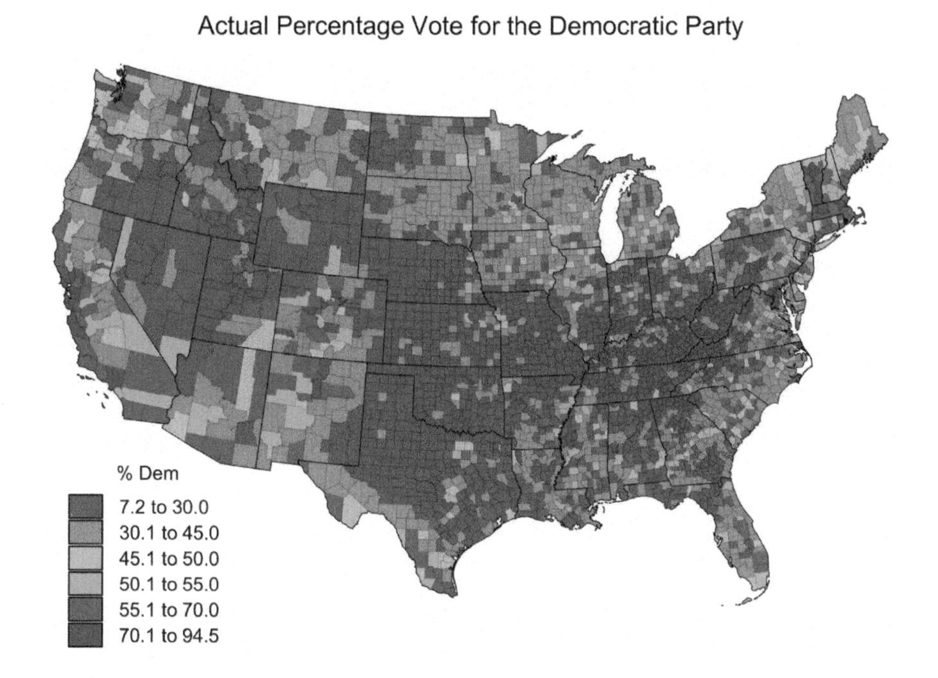

FIGURE 7.1
Percentage Vote for the Democratic Party in the 2020 US Presidential Election by County for the Contiguous United States.

In order to understand the determinants of the percentage vote shown in Figure 7.1, we employ the same model used in Fotheringham et al. (2021) who model the outcome of the 2016 Presidential election and who provide a justification for the model form and the set of explanatory variables included in the model (those in Table 7.2). The socioeconomic covariates were obtained from the United States Census Bureau's American Community Survey (ACS) five-year estimate dataset for 2015–2019. The expectation was that the county-level percentage vote for the Democratic Party would increase with increasing concentrations of Black, Hispanic, and foreign-born populations, and with increasing concentrations of voters with a university education and living in urban areas. Conversely, it was expected that the Democratic share of the vote would decrease as the proportion of elderly voters increased and as median income increased. There were no *a priori* expectations for the direction of the relationships between the Democratic vote and the remaining variables although all of these have been discussed in either the academic literature or the media as possible influences on voter preferences. For instance, an increase in voter turnout could have either a negative or positive impact on the Democratic vote depending on the socio-demographic composition of marginal voters in a county. Further, there was no expectation as to which, if any, of these factors might have a significant spatially varying influence on voter preferences. These are all issues for which there is little or no theoretical guidance and which have largely been ignored in the literature.

To mitigate the sampling bias of modeling percentages based on small populations, counties with a population smaller than 5,000 were not included in the analysis. Also, because the calibration of MGWR is based on the concept of 'borrowing' data from nearby locations, the counties in the noncontiguous states of Hawaii and Alaska were not included in the analysis. This left a total of 2,807 counties for which data on each of the 14 covariates in Table 7.2 along with the percentage voting for the Democratic Party were used in the calibration of both a global (equation 7.2) and local model (equation 7.3). Both the dependent and independent variables in the two models were standardized to have mean of zero and variance of one. This is recommended for any MGWR calibration in order to produce comparable covariate-specific bandwidth estimates (Fotheringham et al., 2017). More details on this can be found in Chapters 5 and 6.

7.4 Global Model Results

Calibration of the global model (equation [7.2]) with the dependent variable defined in equation (7.1) and the covariates given in Table 7.2 for 2,807 counties yielded an R^2 value of 0.70 and an *AICc* of 4,606. The parameter estimates and associated diagnostics are given in Table 7.3, where the symbol '*' denotes a relationship with a p-value ≤ 0.01.

The global model performs reasonably well with an R^2 value (0.7) that most social scientists would deem acceptable. There is almost always a sizeable random element in human behavior, and voting is no exception. There are no multicollinearity issues with the global model as evidenced by the *VIF* values, which are all well below 10 (a common rule of thumb for identifying situations where collinearity might cause a problem in interpreting the results of the model correctly). The results are also generally consistent with expectations: the percentage Democratic vote in counties is positively associated with higher proportions of Black and Hispanic voters, higher proportions of college-educated voters and foreign-born voters, and is also higher in more urbanized counties *ceteris paribus*. Conversely, the percentage Democratic vote declines as median income increases, *ceteris paribus*. Slightly surprising is that the Democratic vote is positively associated with elderly voters (who are typically characterized as being more right-wing in their views than the average voter) and negatively associated with higher proportions of people without health insurance (who would seemingly be better off under a Democratic President than a Republican one). We return to both of these counterintuitive results, which disappear when a local model is calibrated.

Two aspects of voting were added to the model to minimize misspecification bias in the remaining parameter estimates. These were the proportion of votes going to a Third Party candidate (i.e., not to the Democratic or the Republican Parties) and the turnout in each county. Interestingly, higher values on both favored the Democratic Party suggesting that where people did vote for a Third Party candidate (only 1.8% of the electorate did), these were more frequently voters who would have been expected to vote Republican and that higher turnouts in counties tended to favor the Democrats. The remaining covariates in the model, gender ratio, younger voters, income disparities, and the percentage employed in manufacturing appear to have had no real impact on the election, at least globally, despite media hype to the contrary.

However, before accepting these inferences, we need to recognize two problems with the analysis: one concerns the distribution of the residuals from the calibrated model; the other is that the relationships are assumed to be constant over space. Figure 7.2 shows that the spatial distribution of the residuals from the calibrated global model is clearly not random—Moran's I is 0.27 and indicates significant clustering of the residuals at $p \leq 0.01$. This raises alarms because the spatially correlated residuals violate the assumption in OLS that they are independent and identically distributed (i.i.d). As a result, the OLS regression coefficients reported in Table 7.3 may be biased and the type I error rates inflated (Dormann et al., 2007; Kühn, 2007).

The second potential problem is that the model in equation (7.2) assumes that the relationships being modeled are stationary over space. This assumption is questionable and needs to be examined through the calibration of a local model. If the relationships being modeled exhibit spatial heterogeneity, the results from the global model will hide this potentially interesting spatially diversity. We now explore potential spatial heterogeneity in the determinants of voter preferences through the calibration of the model in equation (7.3) by MGWR.

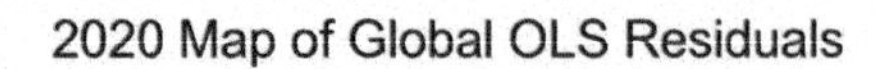

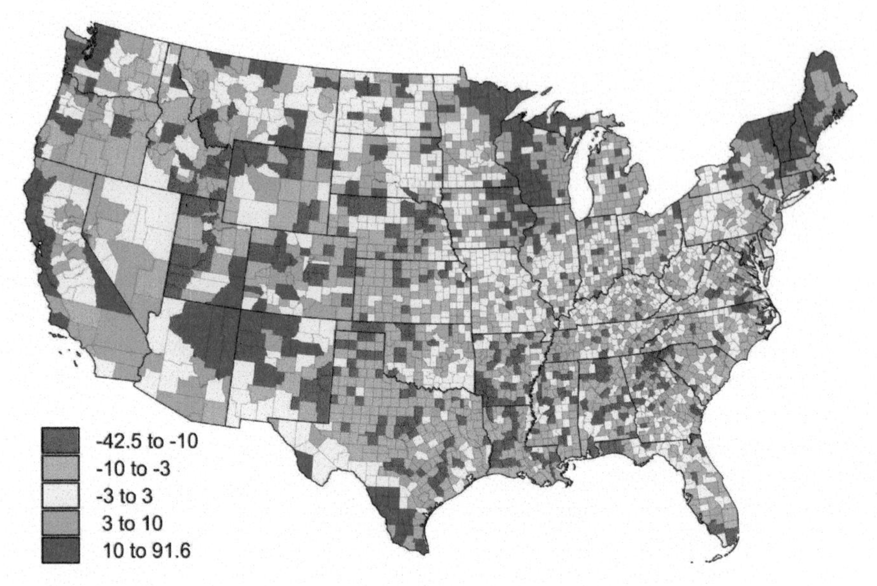

FIGURE 7.2
Spatial Distribution of Residuals From the Global OLS Model.

TABLE 7.3

OLS Calibration Results: Parameter Estimates and Diagnostics.

Covariate	Parameter estimate	S.E.	t	P>|t|	VIF
Intercept	0.000	0.010	0.000	1.000	
sex_ratio	0.018	0.012	1.452	0.147	1.391
pct_black*	0.528	0.013	41.187	0.000	1.538
pct_hisp*	0.272	0.017	15.637	0.000	2.819
pct_bach*	0.450	0.025	17.777	0.000	5.991
med_income*	−0.233	0.024	−9.660	0.000	5.456
pct_65_over*	0.059	0.016	3.575	0.000	2.500
pct_age_18_29	0.036	0.017	2.044	0.041	2.856
Gini	0.017	0.014	1.176	0.240	1.964
pct_manuf	0.006	0.013	0.458	0.647	1.496
ln_pop_den*	0.182	0.016	11.416	0.000	2.379
pct_3rd_party*	0.190	0.013	14.801	0.000	1.543
turn_out*	0.174	0.018	9.900	0.000	2.889
pct_fb*	0.100	0.019	5.360	0.000	3.233
pct_uninsured*	−0.161	0.014	−11.603	0.000	1.792

Source: '*' denotes p-values ≤ 0.01.

7.5 Local Model Results

7.5.1 Diagnostics

The calibration of the model in equation (7.3) by MGWR using MGWR 2.2 (see Chapter 5) on an ordinary laptop with data for 14 covariates for 2,807 spatial units and with the Monte Carlo test for spatial variability of the local parameter estimates turned on took just over six days (six days, one hour, 20 minutes, and 31 seconds).[3] The adjusted R^2 for the model was 0.94, and the *AICc* was 889. Compared to the global equivalents of 0.70 and 4,606, respectively, these values suggest that the local model replicates the patterns of voting in the 2020 US Presidential election significantly more accurately. This is confirmed by both the magnitudes and spatial distribution of the residuals as shown in Figure 7.3. The vast majority of the residuals are in the interval between −3% and 3%, and the Moran's I value is .002 indicating no significant spatial dependency.

Various diagnostics for the local parameter estimates from MGWR are shown in Table 7.4. The optimal bandwidth for each covariate is given in terms of the number of nearest neighbors from which data are drawn (and weighted from 1 to 0 depending on distance from the regression point). Smaller optimal bandwidths indicate processes that vary over space more locally. The 95% confidence intervals for each

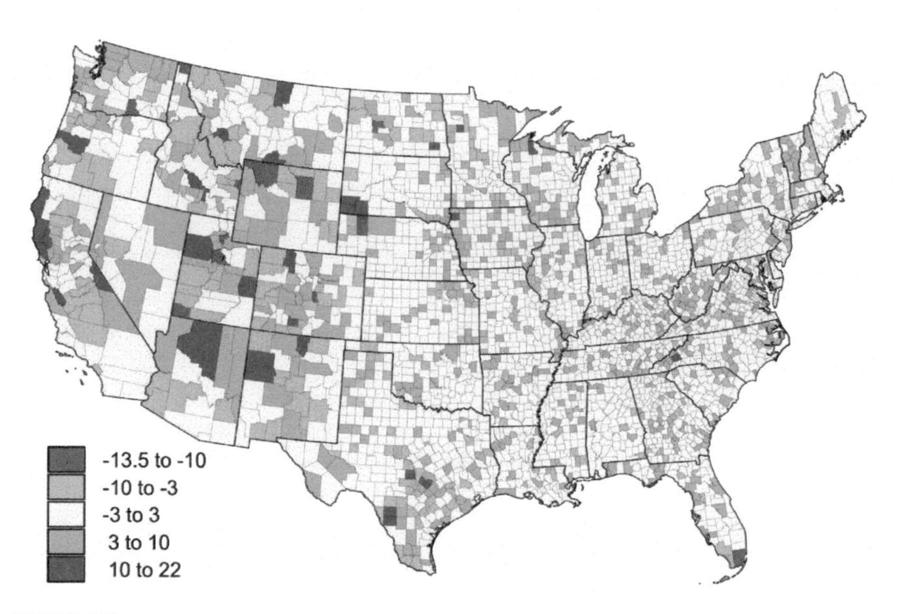

FIGURE 7.3
Spatial Distribution of the Residuals From the MGWR Calibration.

TABLE 7.4

Diagnostics for Local Parameter Estimates Derived From MGWR.

Variable	*BW* (95% CI)	ENP_j	*Adj t-val* (95%)	D_j	*MC p-value*
Intercept	43 (43, 45)	121.4	3.54	0.40	.000*
Sex_ratio	2653 (2154, 2712)	1.5	2.12	0.95	.643
Pct_black	43 (43, 45)	120.7	3.54	0.40	.000*
Pct_hisp	1079 (944, 1348)	3.3	2.43	0.85	.716
Pct_bach	483 (386, 541)	7.3	2.71	0.75	.000*
Median_inc	255 (232, 291)	13.9	2.91	0.67	.020*
Pct_65_over	43 (43, 45)	145.7	3.59	0.37	.006*
Pct_18_29	168 (151, 196)	30.7	3.15	0.57	.056
Gini	503 (445, 600)	10.2	2.82	0.71	.020*
Pct_manuf	196 (173, 232)	24.9	3.09	0.60	.003*
Ln_pop_den	294 (255, 350)	14.2	2.92	0.67	.002*
Pct_3rd_party	1031 (695, 1348)	4.1	2.51	0.82	.913
Turn_out	124 (115, 137)	39.6	3.23	0.54	.005*
Pct_fb	2804 (2404, 2805)	1.2	2.04	0.98	.522
Pct_uninsured	102 (78, 106)	47.6	3.28	0.51	.981

optimized bandwidth are also given. These are not symmetric given the bandwidth is bounded by 43 and *m* (2,807 in this application).

The next three statistics, ENP_j, *Adj t-val(95%),* and *D*, all relate to the optimal bandwidth. ENP_j is the equivalent number of independent parameters and would be 1 if the model were global and *m* if each local regression were independent. *Adj t-val(95%)* represents the adjusted *t*-value for a 95% significance test on each local estimate. This value is adjusted for the issue of multiple hypothesis testing given the degree of dependency between the tests. D_j as defined in Chapter 3, is the degree of dependency within each set of local parameter estimates with values close to zero indicating very low dependency between the estimates (the extreme case would be where the model is calibrated independently for each location) and values close to

1 indicating high dependency between the estimates (the extreme case would be a global process where all the local parameter estimates are the same). The *MC p-value* is the *p*-value derived from the Monte Carlo test for spatial variability and the values designated with an '*' indicate sets of local parameter estimates exhibiting significant spatial variability. It is easy to see the relationship between these diagnostic statistics: relationships exhibiting interesting spatial variability typically have small optimal bandwidths, relatively large values of *ENP* and *Adjusted t*, and smaller values of *D*.

The results in Table 7.4 indicate that there might be considerable spatial heterogeneity in many of the processes being modeled. Specifically, the conditional relationships between voting behavior and the proportion of Black voters, education levels, median income, elderly voters, income disparities, manufacturing employment, population density, and turnout appear to show significant spatial nonstationarity. Other conditional relationships appear to be global. We could crudely categorize the optimized bandwidths in terms of the total number of data points (2,807 here) to indicate processes that vary locally and regionally and which do not vary (i.e., are global). This is done in Table 7.5 by arbitrarily designating local processes as having a bandwidth up to 150, regional processes as having a bandwidth between 150 and 1,999, and global processes as having bandwidths over, or equal to, 2,000.

The next logical stage would be to map the local parameter estimates to get a visual feel for their spatial variability across the United States. However, before proceeding further, we need to check that the model is not picking up any spurious spatial variation in local parameter estimates caused by incorrectly modeling nonlinear relationships with linear functional forms. As described in Chapter 6, this can be done by using the diagnostic test described by Sachdeva et al. (2022), which involves plotting each set of local parameter estimates against the corresponding values of the respective covariate across all 2,807 counties and looking for any relationship in each plot suggesting that the local parameter estimates might be a function of the

TABLE 7.5

Crude Designation of the Spatial Scale Over Which Different Processes Operate.

Local processes (BW $\leq$ 149)	Regional processes (150 $\leq$ BW $\leq$ 1999)	Global processes (BW $\geq$ 2000)
Intercept *	*Pct Hispanic*	*Sex ratio*
*Pct Black**	*Pct. Bach**	*Pct FB*
*Pct 65+**	*Med Inc**	
*Turnout**	*Gini**	
Pct Uninsured	*Pct Manuf**	
	*Ln Pop Den**	
	Pct 3rd Party	
	Pct 18–29	

Source: '*' Indicates Significant Spatial Variation in the Local Parameter Estimates.

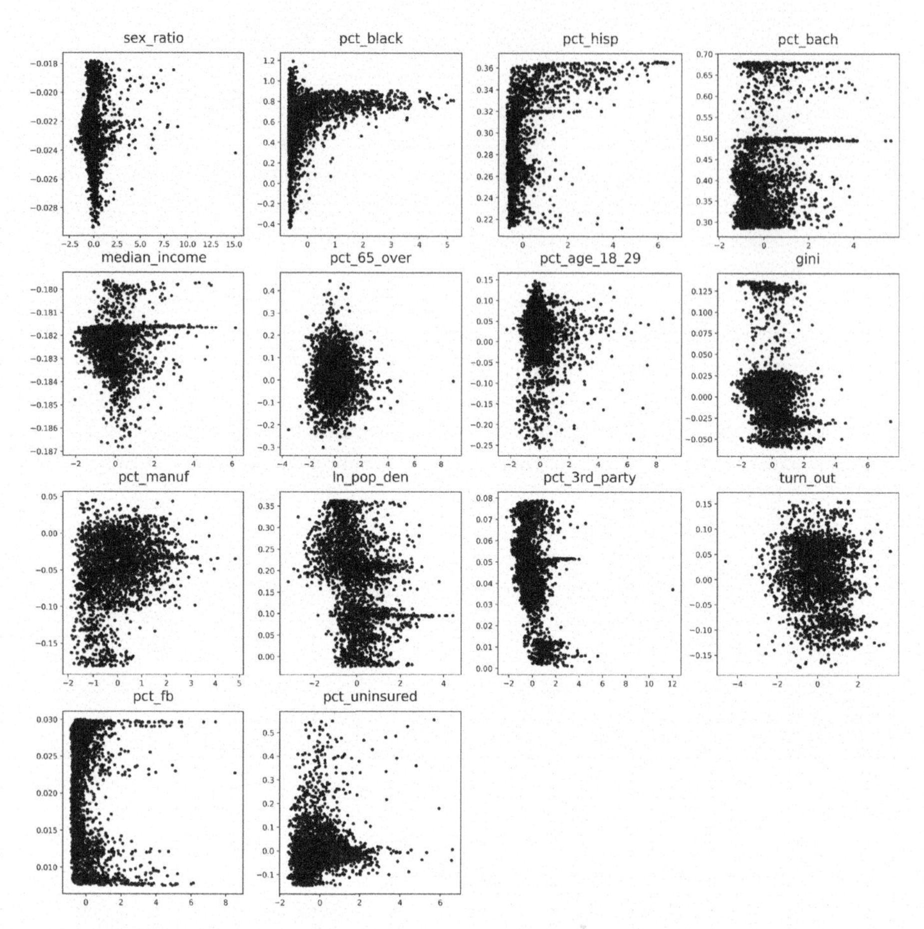

FIGURE 7.4
Checking for Possible Nonlinearities.

covariate magnitude. The plots for each of the 14 covariates in the model are shown in Figure 7.4 with the vertical axis representing the local parameter estimates and the horizontal axis representing the magnitude of the covariate, the latter being in standardized units.

In this case, there appears to be very little structure indicting nonlinearity in any of the plots (for examples of clear structure in such plots, see Figure 6.3 in Chapter 6). In most of the plots the points form amorphous blobs or vertical columns, both suggesting there is no relationship between the magnitudes of the local parameter estimates and the corresponding covariate values. The two plots where there could be some relationship are those for percentage Hispanic population and percentage Black population. The local parameter estimates for the percentage Hispanic population do not exhibit any significant spatial variation, so this is a moot issue. The scatterplot for the percentage Black population comprises two sets of counties (as

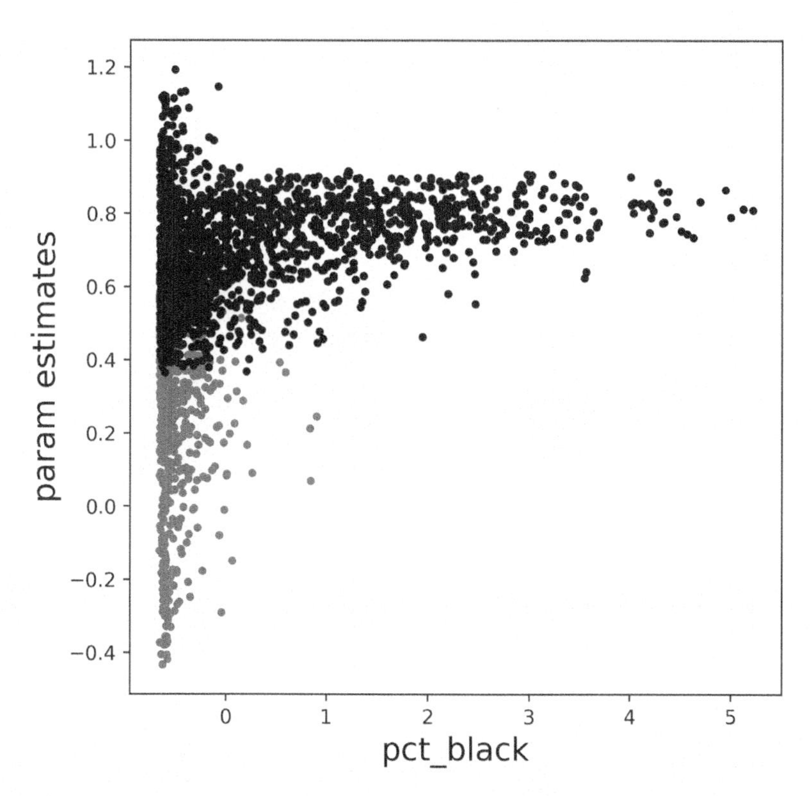

FIGURE 7.5
Scatterplot of the Local Parameter Estimates for the *Pct Black* Covariate Against Levels of the Covariate for each of the 2,807 Counties (gray dots indicate insignificant parameter estimates).

highlighted in Figures 7.5 and 7.6 [a]). In counties where the percentage of Black population is below the national mean (values less than 0 here as the data are standardized), the local parameter estimates for those counties form a vertical column, and many of the lower values are not significant (see Figure 7.5). In counties where the percentage is above the national mean, the local estimates form a horizontal line at around 0.8. So, although there is significant variation in the local parameter estimates for the *Pct Black* covariate, as shown in Figure 7.6, this is because of the presence of sizeable clusters of counties in the United States where the covariate level is almost zero and for which the parameter estimates are insignificant.

7.5.2 Evidence of *Behavioral* Contextual Effects

Given the lack of evidence for any nonlinear relationships in the model, maps of the local parameter estimates exhibiting significant spatial variation (behavioral contextual effects—see Chapter 1) are displayed in Figure 7.6 (a–h) (in addition, intrinsic contextual effects, as measured by the local intercept, also exhibit significant spatial nonstationarity, and these are shown in Figure 7.8 and discussed in Section 7.5.3).

These maps can be explored and contrasted to the single global parameter estimates in Table 7.3 to gain insights into the spatially varying nature of the determinants of how people voted in the 2020 US Presidential election.

Spatial variations in the local parameter estimates relating the proportion of Black population within a county to the percentage votes for the Democratic Party (Figure 7.6 [a])

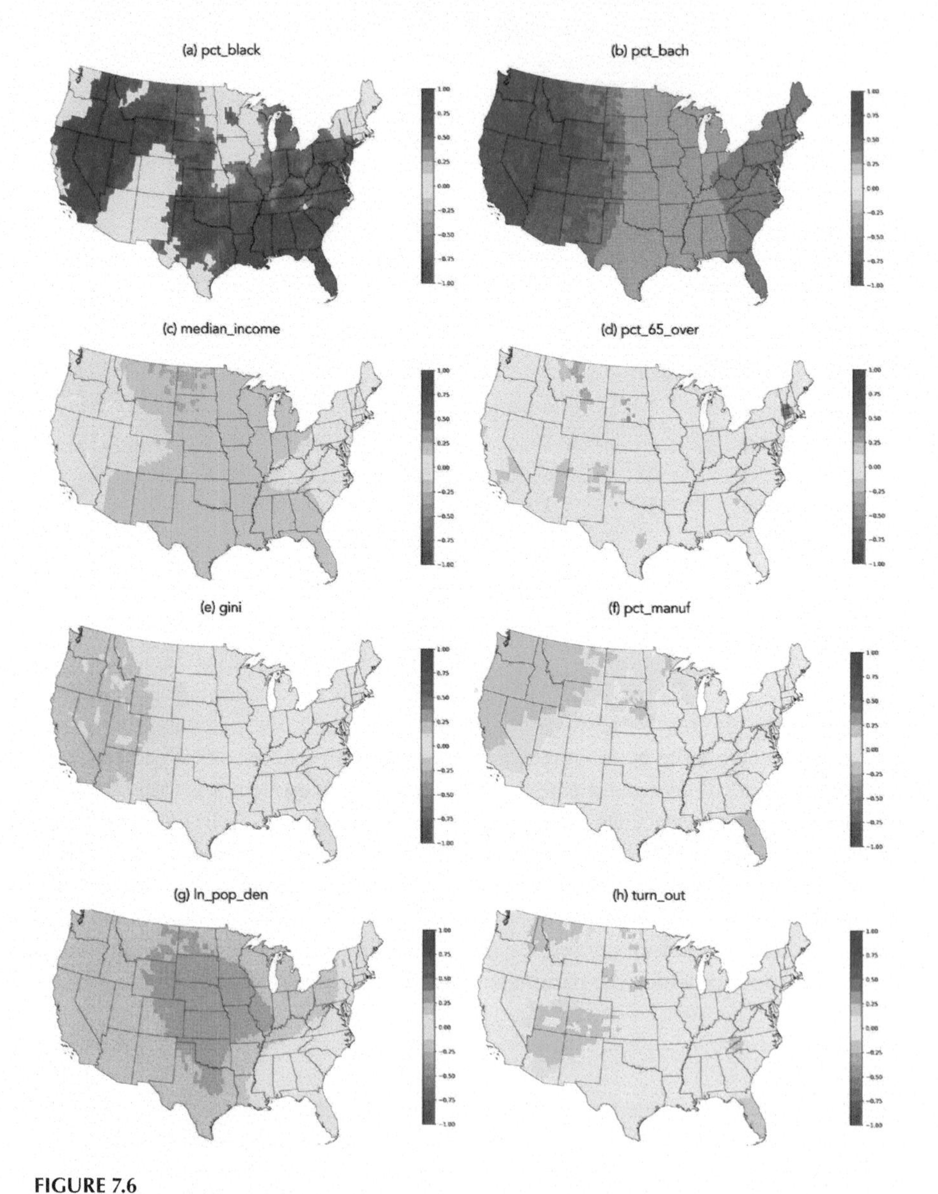

FIGURE 7.6
Local Slope Parameter Estimates Exhibiting Significant Spatial Variability: Behavioral Contextual Effects in Voting Behavior.

show what was alluded to earlier in the examination of possible nonlinear relationships. In counties with very small percentages of Black voters, such as those in Washington, Oregon, Arizona, New Mexico, Colorado, southern Texas, Minnesota, Wisconsin, Iowa, and most of New England, the local parameters are insignificant because the covariate levels are consistently low. Across the rest of the country the influence of the percentage of Black population on votes for the Democratic Party is uniformly strong with parameter estimates typically around 0.8 (the variables are standardized so the parameter estimates can be compared). Interestingly, the global parameter estimate for this covariate is only 0.53 and probably underestimates the impact of Black voters on the election outcome because it incorporates data from counties with very small proportions of Black residents. The percentage of Black voters in a county is the strongest indicator of voting preference of any of the 14 covariates in the model. The optimal bandwidth is 43 (the minimum value possible) indicating a relationship that varies very locally, but again this is largely due to the very low levels of the covariate in many areas rather than the preferences of Black voters being very different across the country.

The impact of education level on voter preference (Figure 7.6 [b]) is uniform in direction (higher educational levels lead to increased votes for the Democratic Party, *ceteris paribus*) across the country but varies in strength regionally (the optimal bandwidth is 483) with the strongest relationships being in the west and the weakest relationships being in the central part of the country. Education is a particularly strong discriminating factor in the way people vote in California, Nevada, Oregon, and Washington with parameter estimates often in excess of 0.75.

The impact of income on voter preference for the Democratic Party is typically negative (Figure 7.6 [c]), but it varies regionally (optimal bandwidth is 255) with the impact being insignificant in most of the northeast, in the eastern seaboard, and in most of California and Nevada. The local model highlights some clear distinctions along state lines such as the boundaries between Ohio and Pennsylvania, Ohio and West Virginia, and Ohio and Kentucky with income being a significant driver of voter preferences in Ohio but not in the other three states.

Figure 7.6 (d) shows the utility in mapping only the significant local parameter estimates. Here, the parameters relate to the percentage of the population aged 65+. and although the Monte Carlo tests suggest that the local estimates exhibit significant spatial variation, the distribution suggests that this variable has little influence on voting preference across the nation. Although there are small pockets of both significantly positive and significantly negative estimates, the vast majority of the local parameter estimates are insignificant. The global parameter estimate, although small (0.06), is significantly positive, which would lead to the incorrect inference that higher proportions of elderly voters in a county favor the Democratic Party.

The local parameter estimates for the Gini coefficient in Figure 7.6 (e) show how income disparities within counties affect voter preferences. Interestingly, the variable is not significant in the global model and locally it is not significant throughout most of the country, yet it is significantly negative across the western US and significantly positive in much of Florida. This suggests that voters exposed to greater income inequalities have a slight preference for the Democratic Party in the western United States, *ceteris paribus*, but favor the Republican Party in most of Florida, *ceteris paribus*.

Manufacturing employment (Figure 7.6 [f]) has long been linked to votes for the Democratic Party, certainly across much of the northeast, yet it has an insignificant effect in the global model and throughout most of the country, including the northeast, according to the local model. However, there is a significant negative relationship with voting for the Democratic Party across most of the northwest and in the state of Florida. Interestingly, the local model picks out the Florida border with Alabama and Georgia despite this not being a feature of the model.

Voters in urban areas tend to favor the Democratic Party over the Republican Party, *ceteris paribus*. This is confirmed by both the global model result and the local parameter estimates as shown in Figure 7.6 (g), which are significantly positive across most of the country. However, the local estimates show that in much of the southeast (except Florida) there is no significant urban/rural divide in how people vote, *ceteris paribus*. Again, the local model results highlight state borders remarkably well as evidenced by the difference in significance across the Louisiana/Mississippi border, the Virginia/North Carolina border, and the Kentucky/Tennessee border.

The spatial variation in the effect of turnout is shown in Figure 7.6 (h). The global parameter estimate suggests that high turnout favors the Democratic Party across the country, but the local parameters suggest that this is not the case. Across most of the country turnout had little influence on the outcome of the county-level vote and only in Florida did high turnouts help the Democratic Party. High turnouts actually favored the Republican Party in Arizona, Utah, Colorado, and parts of New Mexico.

Before turning to intrinsic contextual effects, there is one further piece of evidence for behavioral contextual effects, which is shown in Figure 7.7, where the local parameter estimates for the covariate *pct uninsured* are displayed. Although the

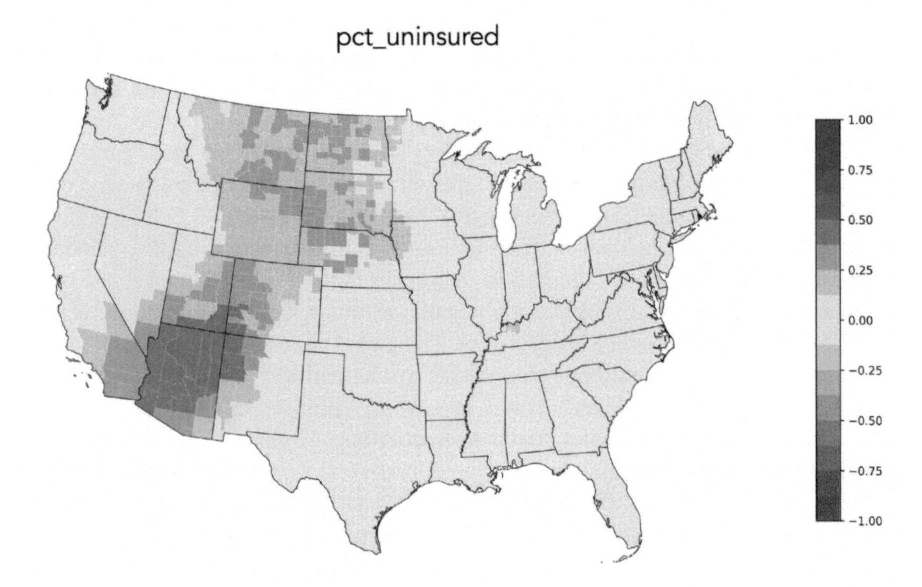

FIGURE 7.7
Distribution of the Local Estimates for the Covariate *pct uninsured*.

Monte Carlo test suggests these do not exhibit any significant spatial variability, the optimal bandwidth is only 102 suggesting a locally varying process. The distribution of the local estimates hints at why neither diagnostic on its own is sufficient in this case. The pattern is one where throughout much of the country the effect of having no health insurance on voter preference is insignificant (perhaps explaining why the Monte Carlo test for spatial variability is insignificant) but in a broad swathe of counties running from Arizona in the southwest to Montana and North Dakota in the north, not having health insurance significantly swayed voter preferences toward the Democratic Party (perhaps explaining the low optimal bandwidth). Such nuances are completely missed in the global model calibration, which counterintuitively suggests that not having health insurance has a significantly *negative* effect on voting for the Democratic Party across the country. Fotheringham and Sachdeva (2022) show how the differences between global and local model calibrations can produce sign reversals leading to instances of a spatial variant of Simpson's paradox.

7.5.3 Evidence of *Intrinsic* Contextual Effects

The previous section details the evidence for the presence of behavioral contextual effects—where location affects the intensity with which various socioeconomic determinants affect behavior. We now turn to intrinsic contextual effects, which are arguably more important and which relate to the direct effect of place on behavior.

The ability to estimate a local intercept in MGWR allows the separation of intrinsic contextual effects and socioeconomic effects on voter preferences. It further allows the influence of both of these contributions to the way people vote to be quantified. To see this, consider the local model in equation (7.3) rewritten as

$$y_i^* = \alpha_i + \sum_k \beta_{ik} x_{ik}^* + \varepsilon_i \tag{7.4}$$

where y_i^* is the standardized value of y_i, the percentage vote for the Democratic Party, for county i, x_{ik}^* represents the standardized value of the kth covariate x for county i, α_i is the local intercept, and β_{ik} represents the slope coefficient for the kth covariate for county i. This equation can be rewritten as

$$\frac{(y_i - \bar{y})}{\sigma_y} = \alpha_i + \sum_k \beta_{ik} \frac{(x_{ik} - \bar{x}_k)}{\sigma_{x_k}} + \varepsilon_i \tag{7.5}$$

where $\bar{y}$ denotes the mean county value of y_i, $\bar{x}_k$ denotes the mean of the kth covariate over all the counties, and σ represents a standard deviation. This equation can be expanded and rearranged to produce an expression for the percentage vote for the Democratic Party in each county, y_i, as

$$y_i = \bar{y} + \alpha_i \sigma_y + \sigma_y \sum_k \beta_{ik} \frac{(x_{ik} - \bar{x}_k)}{\sigma_{x_k}} + \varepsilon_i \sigma_y \tag{7.6}$$

which has three operational components. The term $\bar{y}$ is the mean vote for the Democratic Party across all counties, $\alpha_i\sigma_y$ is the proportion of the vote for the Democratic Party due to location or 'context', and $\sigma_y \sum_k \beta_{ik} \frac{(x_{ik} - \bar{x}_k)}{\sigma_{x_k}}$ is the proportion of the vote for the Democratic Party due to the mix of population within each county. Here, we are interested in the spatial distribution of the term $\alpha_i\sigma_y$, which denotes the role of intrinsic contextual effects in determining the percentage vote for the Democratic Party. Although we do not know each α_i, we have estimates of these values from the calibrated model which are shown in Figure 7.8. These describe the influence of location (place) on voting behavior: counties shaded blue indicate where location has a significant positive influence on votes for the Democratic Party; counties shaded red are places where location has a significant negative influence on voting for the Democratic Party. Counties shaded gray are where there is no significant effect of location and represent parts of the country where voters are intrinsically neutral in their political leanings, once socioeconomic factors are accounted for. It is important to note that Figure 7.8 displays the *intrinsic* leanings of voters to either the Democratic (blue) or Republican (red) Party. It does *not* display how counties actually voted because the impact of the socioeconomic composition of each county has been subtracted.

These local estimates of the intercept can be directly translated into voting influence for the Democratic Party by multiplying each value by σ_y, which is simply the standard deviation of the unstandardized percentage votes for the Democratic Party in each county. This results in the distribution in Figure 7.9, which shows the

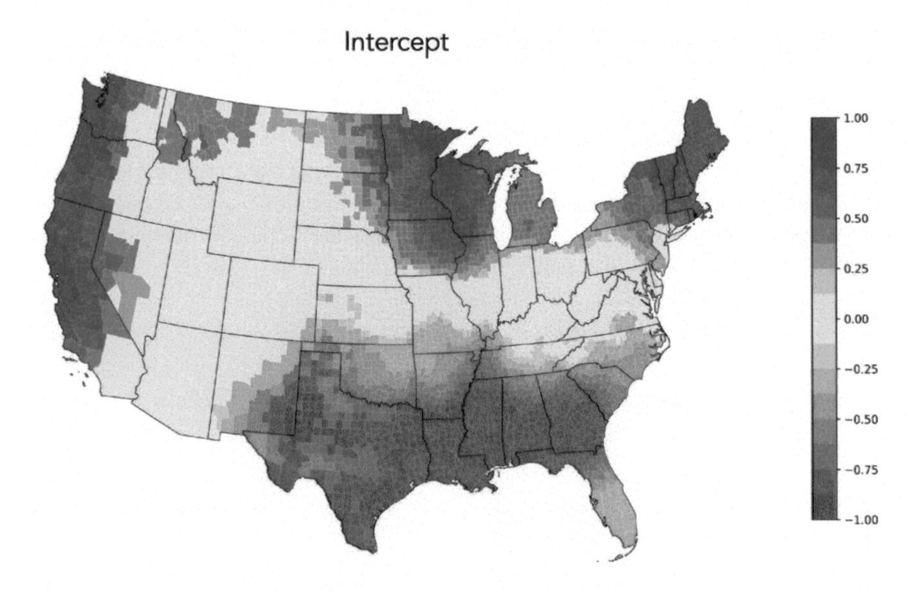

FIGURE 7.8
Local Intercept Estimates: Intrinsic Contextual Effects in Voting Behavior (*insignificant estimates are in gray*).

percentage changes in vote share for the Democratic Party due to intrinsic contextual effects. Context was responsible for *increasing* the Democratic share of the vote by over 10 percentage points in the most intrinsically Democratic counties (those shaded dark blue) and was responsible for *reducing* the Democratic vote by over 10 percentage points in the most intrinsically Republican counties (those in dark red).

The distribution in Figure 7.9 makes a powerful case for geographical context because it shows that even if all counties had exactly the same socioeconomic mix of voters, votes would not be cast in the same way throughout the country due to spatially varying contextual effects. Another way of saying this is that if two people with exactly the same socioeconomic profiles grew up in different parts of the country, one, say, in western Oregon and the other in Alabama, it is very unlikely they would share the same preference for the Democratic Party due to the unseen influence of geographical context. This influence has long been recognized (Taylor, 1973; Johnston et al., 1990; Agnew, 1996; Sui & Hugil, 2002; Warf & Leib, 2011; Agnew, 2014; Lappie & Marschall, 2018) but until recently, it has never been quantified.

The map in Figure 7.9 is also important in describing and quantifying the deep and persistent divisions in US political leaning between the north and the south. For example, the extent of the intrinsically Republican leaning counties is amazingly coincidental with those supporting the Confederacy in the Civil War, which took place over 150 years ago! For further details on the this and also on the persistence

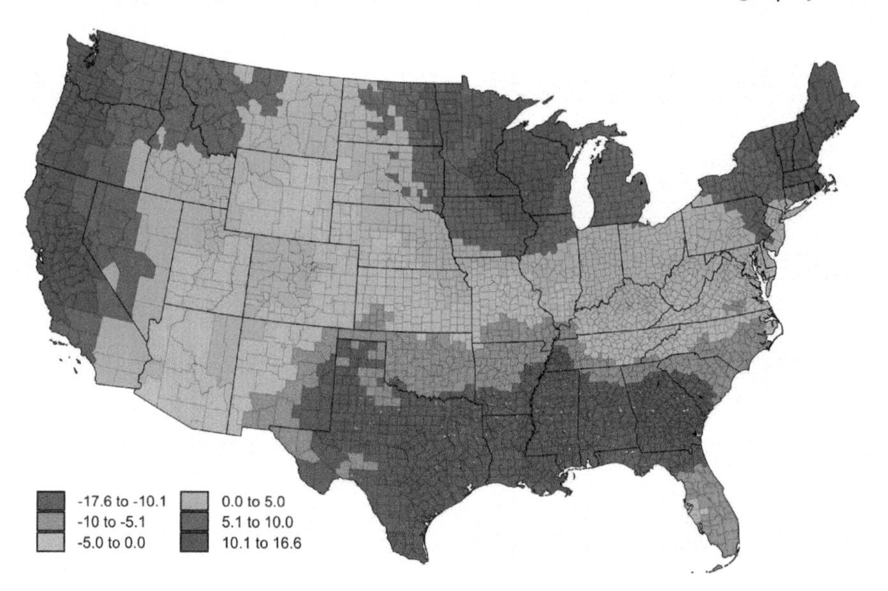

FIGURE 7.9
Percentage Contribution to the Democratic Party Vote in Each County Due to Intrinsic Contextual Effects.

of these trends through four different elections, see Fotheringham et al. (2021) and Li and Fotheringham (2022).

We can also measure how much of each county's vote can be attributable to its socioeconomic profile by calculating the expression $\sigma_y \sum_k \beta_{ik} \frac{(x_{ik} - \bar{x}_k)}{\sigma_{x_k}}$ in equation (7.6), and these values are shown in Figure 7.10. These are percentages that are either subtracted or added to the vote for the Democratic Party due to the particular socioeconomic mix of voters in each county. The counties shaded in red are where the socioeconomic composition of the population reduces the Democratic vote; those shaded blue are where the population composition increases the Democratic vote. Interestingly, the spatial distribution of the impacts of population composition on voting for the Democratic Party is almost the opposite of the impacts of the contextual effects shown in Figure 7.9. Across much of the south, counties should favor the Democratic Party based on their population composition (large proportions of Black and/or Hispanic voters and lower median incomes) while across much of the north, counties should favor the Republican Party (largely white, affluent voters).

The values in each county from Figures 7.9 and 7.10 can be added to the mean vote for the Democratic Party across the 2,807 counties to provide the MGWR estimate of the actual vote in each county. These predicted values are shown in Figure 7.11 left in juxtaposition with the actual percentage vote in Figure 7.11 right.

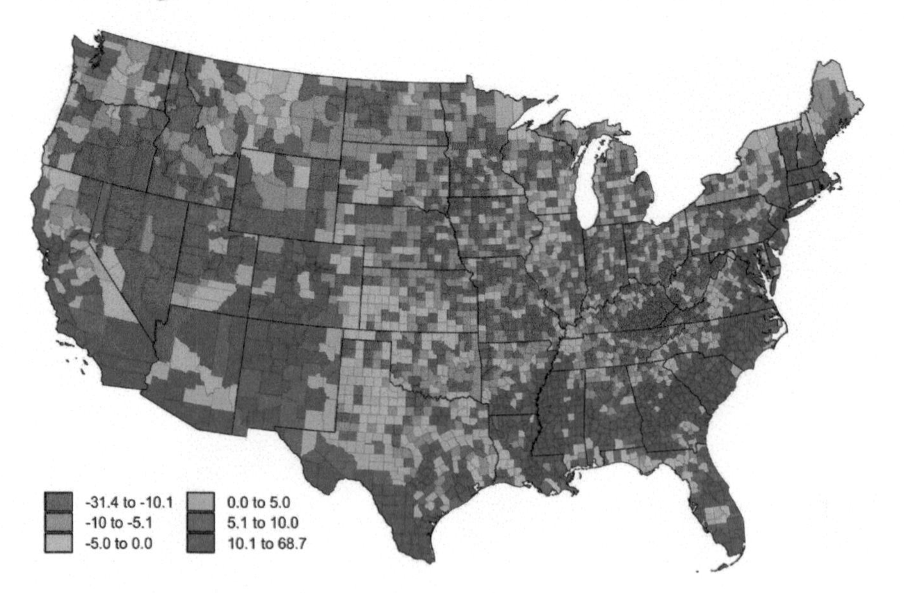

FIGURE 7.10
Percentage Contribution to the Democratic Party Vote Caused by the Socioeconomic Mix of Population in Each County.

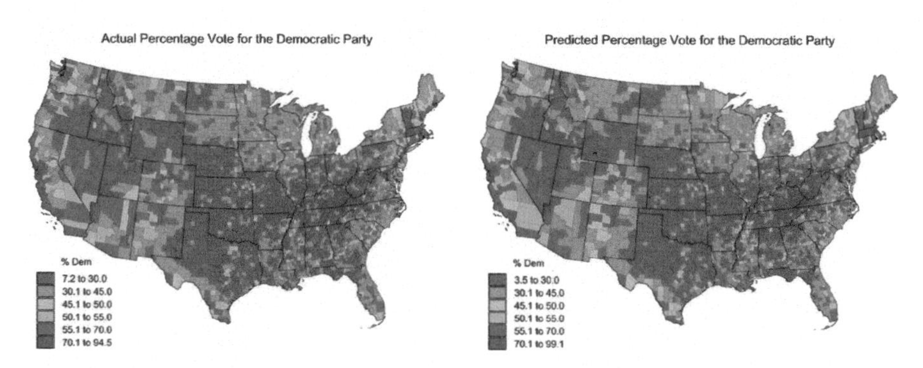

FIGURE 7.11
Actual and Predicted Percentage Votes for the Democratic Party by County in the 2020 Election.

7.6 Discussion

This chapter serves two purposes: (i) to demonstrate the insights that can be gained from the calibration of local models and (ii) to act as a template of good practice for local modeling. We now briefly discuss both of these.

In terms of insights, the application of MGWR to the investigation of voter preferences strongly suggests that some of the factors influencing voting behavior are not constant over space and therefore global models of voting behavior are inappropriate in modeling such behavior. Local models such as MGWR provide information not just on how the determinants of voter preferences vary over space but also on the spatial scale of this variation. Such information is hidden behind the 'average' estimates of behavior that are produced by global models. In addition, through its locally varying intercept, MGWR quantifies the degree to which voter preferences are determined by intrinsic contextual effects, independent of the socioeconomic determinants of such preferences. Not only does this suggests that where a person lives is a major factor in how that person will vote, irrespective of their socioeconomic background, but it also validates and quantifies place-based geography. That a model calibrated in one location does not work well in another location should therefore not be a surprise or a concern—the processes being modeled may well vary over space. Models that do not explicitly account for local contextual effects and spatially varying preferences and values are at risk of producing poor estimates of behavior and also biased estimates of the parameters associated with the other determinants of such behavior.

In terms of good practice in local modeling, we stated earlier that no empirical application is perfect and that often many decisions have to be taken in the calibration of a model, which can be the subject of criticism or debate. Local modeling is no exception to this as it is relatively easy to produce impressive-looking results that are essentially noise. However, certain steps can and should be taken to avoid some obvious pitfalls, and in Chapter 6 we provided a checklist for this purpose, which we repeat here. It is useful to consider each of these items in the context of the empirical application described earlier.

1. ***Does the application of a local model make sense?*** By this we mean if the inference from the model calibration suggests that the processes being modeled are spatially varying, can such variation be explained? In the case of voting behavior, it makes sense that location (context) could play a role in determining people's preferences for one political party over another, and there is a large literature that supports such a belief. The local model results not only provide additional support for the role of geographical context but also quantify its importance in determining the outcomes of elections.
2. ***Do the data, in terms of both the number of observations and their spatial distribution, allow a sensible model calibration?*** Yes. There are 2,807 contiguous observations, which are in a relatively compact boundary (counties in the mainland United States).
3. ***Is the global model defensible in terms of both the variables included and their functional form? Have appropriate checks been undertaken to ensure this?*** Yes, the global model R^2 value is 0.7, which is fairly robust for a social science application. The model contains all the variables that the literature suggests might affect a voter's political preference. Other variables were considered but were found to have little value. The functional form of each relationship was examined, and a natural logarithmic transformation was made to the population density variable.
4. ***Has a check been undertaken to ensure that nonlinearity is not a possible cause of spatially varying parameter estimates rather than spatially varying processes?*** Yes. The diagnostic analysis described by Sachdeva et al. (2022) was undertaken to check if nonlinear relationships could be a source of parameter spatial nonstationarity.
5. ***Have all the data been standardized prior to calibration?*** Yes. This allows comparisons of the bandwidths and the parameter estimates.
6. ***Have suitable inferential tests been undertaken on both the individual local estimates and also the local parameter surfaces? The former should include corrections for multiple hypothesis testing and dependency.*** Yes. Inferential tests incorporating both multiple hypothesis testing and degree of dependency were undertaken on each local parameter estimate and also on each set of local estimates for spatial variability, the latter using a Monte Carlo procedure.
7. ***If the local parameter estimates are depicted on a map, have nonsignificant estimates been suitably identified?*** Yes, nonsignificant parameter estimates were depicted in gray when mapped.

Notes

1. It might be tempting to claim that the prevalence of Evangelical Christianity is such a variable because Evangelicals tend to be right-wing in their views (and therefore tend to favor the Republican Party, particularly when led by Trump), and there is a large difference in the adherence to this type of faith between the two counties (Iowa Co. 2.9%; Panola Co. 33.6%). However, the reasons people favor Evangelical Christianity are very similar to why they vote Republican, so adding this variable to the voting model would explain very little, if anything, about the determinants of voting behavior, and it would certainly create severe misspecification bias in the remaining parameter estimates. One could equally have a model with the same set of covariates as in Table 7.2 in which the dependent variable was the proportion of Evangelical adherents in each county, and context would also play a significant role in this model.
2. The data were retrieved from https://electionlab.mit.edu
3. The MC test for the spatial variability of the local parameter estimates essentially calibrates 1,000 separate MGWR models, so the run time is massively decreased when this option is not turned on. Without this option turned on, the calibration took approximately one hour on the same machine.

References

Agnew, J. (1996). Mapping politics: How context counts in electoral geography. *Political Geography*, *15*(2), 129–146.

Agnew, J. (2014). *Place and politics: The geographical mediation of state and society*. Milton Park: Routledge.

Dormann, C. F., McPherson, J. M., Araújo, M. B., Bivand, R., Bolliger, J., Carl, G., Davies, R. G., Hirzel, A., Jetz, W., Kissling, W. D., Kühn, I., & Ohlemuller, R. (2007). Methods to account for spatial autocorrelation in the analysis of species distributional data: A review. *Ecography*, *30*(5), 609–628.

Fotheringham, A. S., Li, Z., & Wolf, L. J. (2021). Scale, Context and heterogeneity: A spatial analytical perspective on the 2016 US Presidential election. *Annals of the American Association of Geographers*, *111*(6), 1602–1621.

Fotheringham, A. S., & Sachdeva, M. (2022). Scale and local modeling: New perspectives on the modifiable areal unit problem and Simpson's paradox. *Journal of Geographical Systems*, *24*(3), 475–499.

Fotheringham, A. S., Yang, W., & Kang, W. (2017). Multiscale geographically weighted regression (MGWR). *Annals of the American Association of Geographers*, *107*(6), 1247–1265.

Hillygus, D. S., & Jackman, S. (2003). Voter decision making in election 2000: Campaign effects, partisan activation, and the Clinton legacy. *American Journal of Political Science*, *47*(4), 583–596.

Johnston, R., Shelley, F. M., & Taylor, P. J. (1990). *Developments in electoral geography*. Milton Park: Routledge.

Kim, J., Elliott, E., & Wang, D. M. (2003). A spatial analysis of county-level outcomes in US Presidential elections: 1988–2000. *Electoral Studies*, *22*(4), 741–761.

Kühn, I. (2007). Incorporating spatial autocorrelation may invert observed patterns. *Diversity and Distributions*, *13*(1), 66–69.

Lappie, J., & Marschall, M. (2018). Place and participation in local elections. *Political Geography*, *64*, 33–42.

Leigh, A. (2005). Economic voting and electoral behavior: How do individual, local, and national factors affect the partisan choice? *Economics & Politics*, *17*(2), 265–296.

Levernier, W., & Barilla, A. G. (2006). The effect of region, demographics, and economic characteristics on county-level voting patterns in the 2000 Presidential election. *Review of Regional Studies*, *36*(3), 427–447.

Li, Z., & Fotheringham, A. S. (2022). The spatial and temporal dynamics of voter preference determinants in four U.S. Presidential elections (2008–2020). *Transactions in GIS*, *26*(3), 1609–1628.

Mutz, D. C. (2018). Status threat, not economic hardship, explains the 2016 Presidential vote. *Proceedings of the National Academy of Sciences*, *115*(19).

Powell, G. B. (1986). American voter turnout in comparative perspective. *American Political Science Review*, *80*(1), 17–43.

Sachdeva, M., Fotheringham, A. S., Li, Z., & Yu, H. (2022). Are we modelling spatially varying processes or non-linear relationships? *Geographical Analysis*, *54*(4), 715–738.

Scala, D. J., Johnson, K. M., & Rogers, L. T. (2015). Red rural, blue rural? Presidential voting patterns in a changing rural America. *Political Geography*, *48*, 108–118.

Schaffner, B. F., MacWilliams, M., & Nteta, T. (2018). Understanding white polarization in the 2016 vote for President: The sobering role of racism and sexism. *Political Science Quarterly*, *133*(1), 9–34.

Sigelman, L., & Sigelman, C. K. (1982). Sexism, racism, and ageism in voting behavior: An experimental analysis. *Social Psychology Quarterly*, *45*(4), 263–269.

Sui, D. Z., & Hugill, P. J. (2002). A GIS-based spatial analysis on neighborhood effects and voter turn-out: A case study in College Station, Texas. *Political Geography*, *21*(2), 159–173.

Taylor, P. J. (1973). Some implications of the spatial organization of elections. *Transactions of the Institute of British Geographers*, *60*, 121–136.

Warf, B., & Leib, J. (2011). *Revitalizing electoral geography*. Burlington: Ashgate.

8

MGWR and Other Models Incorporating Spatial Contextual Effects

8.1 Introduction

In previous chapters, we have shown how MGWR can model both intrinsic and behavioral contextual effects through estimates of the local intercept and local slope parameters, respectively. There are other modeling frameworks that can take contextual effects into account, and it is useful to compare these frameworks to MGWR. One class of models that incorporate spatial contextual effects are spatial econometric models, the most common of which are the spatial lag model (SLM) and the spatial error model (SEM) (Anselin, 1988; Anselin & Bera, 1998; LeSage & Pace, 2009). These models are primarily used to remove spatial autocorrelation in the residuals by using a spatial autoregressive process on either the outcomes (SLM) or the errors (SEM). The SLM cannot be directly compared with MGWR because the spatial interaction effects between observations in the SLM model are not the same as the contextual effects in the MGWR model. However, recent studies have proposed hybrid models that combine MGWR with SLM, more details of which can be found in Chen et al. (2022) although it is not clear why one would want to have two components accounting for error covariance. The SEM can be considered as a special case of MGWR where the filtered autoregressive residuals have a similar effect to the local intercept in MGWR, capturing locational influences that are omitted or misspecified in the model. However, the SEM model only includes global parameter estimates, so the effects associated with covariates are not allowed to vary spatially. Consequently, while a SEM can capture intrinsic contextual effects, it cannot capture behavioral contextual effects (see Chapter 1 for details on intrinsic and behavioral contextual effects). Furthermore, the SEM is conditioned on the spatial weight matrix specification, which is often arbitrary, and it does not allow for inference on the intrinsic contextual effects.

Another framework that can capture contextual effects, which is not strictly spatial but has been widely applied to geographic data, is that of multilevel modeling (MLM) (also known as mixed modeling or hierarchical linear modeling). Geographic data are often multileveled: examples include children within school districts, houses within neighborhoods, and counties within states. Multilevel models acknowledge that there might be heterogeneity in relationships between levels

DOI: 10.1201/9781003435464-8

of the hierarchy, which can be modeled by so-called random effects. Examples of multilevel model applications in geographical studies include modeling the health outcomes of individuals exposed to environmental effects (Duncan et al., 1998; Zahnd & McLafferty, 2017; Ma et al., 2018), measuring neighborhood effects of house prices (Orford, 2000; Dong et al., 2015), and estimating small-area statistics by combing aggregated and survey data (Twigg et al., 2000; Park et al., 2004). Multilevel models are not able to capture intrinsic and behavioral contextual effects at the individual level but can model these effects at an aggregated or higher level. However, the aggregated level needs to be defined *a priori*, which is not always possible, and may be subject to the modifiable area unit problem (MAUP) if the underlying processes operate at different spatial scales to those defined *a priori*.

In this chapter, we compare MGWR with SEM and MLM using Monte Carlo simulations and empirical data. We show that all three frameworks can produce similar results regarding the role of intrinsic spatial context although only MGWR can incorporate the role of behavioral spatial context accurately. More specifically, in Section 8.2, we introduce the basics for spatial error and multilevel models. Section 8.3 presents a comparison of the two models with MGWR based on Monte Carlo simulations. In Section 8.4, we apply all three methods to the election dataset described in Chapter 7. As machine learning has increasingly been used to model spatial data as an alternative to spatial models, in Section 8.5 we include a comparison between MGWR and machine learning for the simulated dataset. Section 8.6 concludes the chapter with some further thoughts.

8.2 Spatial Error Models and Multilevel Models

8.2.1 Spatial Error Models

The spatial error model explicitly accounts for any spatial dependence in the errors by spatially filtering the error term using a spatial autoregressive component (Anselin, 1988). The SEM is formulated as

$$y = \boldsymbol{X\beta} + \boldsymbol{u} \tag{8.1}$$

$$\boldsymbol{u} = \lambda \boldsymbol{W u} + \varepsilon \tag{8.2}$$

where $\boldsymbol{u}$ is the unfiltered error term, λ is the parameter for the spatial autoregressive term, $\boldsymbol{W}$ is an n-by-n spatial weights matrix, and ε is the remaining random error. The parameter λ measures the sign and the magnitude of the spatial dependency in the error term. When λ is zero, the regression function reduces to ordinary least squares (OLS). The weight matrix $\boldsymbol{W}$ can be specified *a priori* in many different ways, which adds an element of subjectivity to the model calibration (see Yu & Fotheringham, 2022, for examples of the dependency of measures of spatial dependency on the definition of the spatial weights matrix). Common examples of *a priori* definitions include those based on Queen or Rook contiguities, k nearest neighbors, and a fixed distance band with or without a decay function. Alternatively, the spatial weights matrix can be selected using a data-driven process to find the appropriate specification that optimizes a model selection criterion such as *AIC* or *BIC* (Chi & Zhu, 2019). Consequently, the term $\lambda \boldsymbol{Wu}$ is the spatial autoregressive error term centered at zero and exhibiting a certain spatial pattern conditional on the strength of the error autocorrelation and the spatial weights matrix. By combining equations (8.1) and (8.2), we can see in (8.3) that $\lambda \boldsymbol{Wu} + \boldsymbol{\beta}_0$ serves a similar role to the local intercept vector $\boldsymbol{\beta}_0$ in an MGWR model, which, as shown in earlier chapters, can be interpreted as an intrinsic contextual effect that is independent of the compositional effect. However, SEM is not able to capture behavioral contextual effects because the rest of the equation remains the same for all locations.

$$\boldsymbol{y} = \left(\lambda \boldsymbol{Wu} + \boldsymbol{\beta}_0\right) + \boldsymbol{X}\boldsymbol{\beta}_{\neq 0} + \boldsymbol{\varepsilon} \tag{8.3}$$

8.2.2 Multilevel Models

Multilevel models are appropriate for hierarchically structured data, a data type quite commonly found in geographical analysis. For simplicity and comparability to MGWR, we present a two-level model, with individuals and groups, with only two covariates, but more complex multilevel models can be constructed in similar ways with many nested hierarchical levels. The level 1 (individuals) regression model is

$$y_{ip} = \beta_{0p} + \beta_{1p}x_{1ip} + \beta_{2p}x_{2ip} + \varepsilon_{ip} \tag{8.4}$$

where y_{ip} is the dependent variable for observation i that belongs to a second-level group p, β_{0p} is the intercept term for group p, x_{1ip} and x_{2ip} are the covariate values for observation i in group p, β_{1p} and β_{2p} are the slopes for group p, and ε_{ip} is the random error. The intercept and slope parameters can vary across the second-level groups, which are shown as

$$\beta_{0p} = \beta_0 + \mu_{0p} \tag{8.5}$$

$$\beta_{1p} = \beta_1 + \mu_{1p} \tag{8.6}$$

$$\beta_{2p} = \beta_2 + \mu_{2p} \tag{8.7}$$

where β_0 is the overall global intercept parameter and μ_{0p} is the random effect meaning the deviation of the intercept of group p from the overall intercept. Similarly, β_1 and β_2 are the overall global slope parameters, and μ_{1p} and μ_{2p} are their deviations from the overall effects. In a standard multilevel model, each random effect follows a normal distribution with mean 0 and an unknown variance, and the estimated variance indicates the magnitude of the between-group heterogeneity. Consequently, spatial contextual effects are represented in the model by variations in the parameter

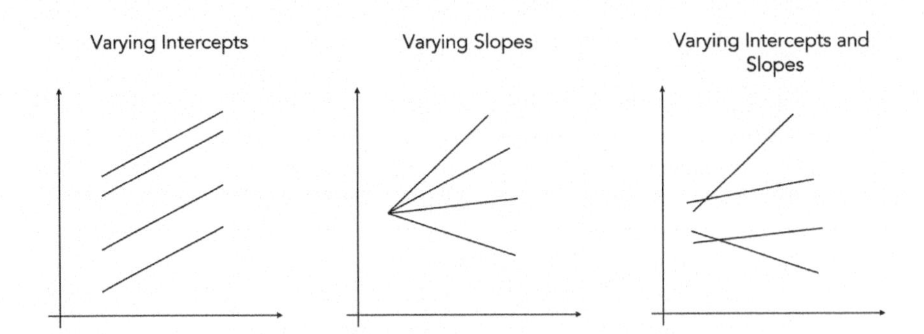

FIGURE 8.1
Illustration of Three Types of Multilevel Model.

estimates across the groups. As shown in Figure 8.1, there are three types of MLM. When both $\mu_{1p} = 0$ and $\mu_{2p} = 0$, this is termed a varying intercepts model, and it can only account for intrinsic contextual effects. When only $\mu_{0p} = 0$, this is a varying slopes model, which allows slopes to vary across the groups. When all the random effects are nonzero, this is a varying intercepts and slopes model, which is the most flexible specification that can account for both intrinsic and behavioral contextual effects. Of course, the degree to which contextual effects can be described is constrained by the *a priori* definition of the groups. When all the random effects are zero, the multilevel model reduces to a linear regression model.

8.3 Comparisons Between Models Using Simulated Data

8.3.1 Monte Carlo Simulation Design

In order to compare the behavior of MGWR, SEM, and MLM models in terms of their ability to capture spatial contextual effects, three spatially varying processes β_0, β_1, β_2 operating at different spatial scales were simulated across a 40-by-40 grid yielding a total of 1,600 observations. Processes β_0 and β_1 are Gaussian random fields $GRF(2, \Omega)$ with mean of 2 and covariance of Ω, which is denoted as

$$\Omega(\boldsymbol{h}) = exp\left(-0.5 * (\boldsymbol{d} / h)^2\right) \tag{8.8}$$

where $\boldsymbol{d}$ is an n-by-n matrix containing pairwise distances for all locations, and h is a scale parameter indicating the amount of distance-decay in the covariance function. Process β_0 is generated with $h = 10$ and operates at a local scale, process β_1 is simulated with $h = 25$ yielding regional spatial variation, and process β_2 is constant with mean 2 and no spatial variation, representing a global process. The GRF surfaces were constructed using the *gstools* python package (Müller et al., 2022) and are shown in Figure 8.2.

A model is then specified as

$$y_i = \beta_{0i} + \beta_{1i} x_{1i} + \beta_{2i} x_{2i} + \varepsilon_i \tag{8.9}$$

where spatial variation in β_{0i} would indicate intrinsic contextual effects and spatial variation in β_{1i} and β_{2i} would indicate behavioral contextual effects. Both covariates and the errors were drawn from a standard normal distribution N (0, 1). For the Monte Carlo simulation, 1,000 realizations of the covariates and error terms were generated, and for each realization the dependent variable is reconstructed according to the model in equation (8.9). An MGWR model and a SEM model were calibrated based on the simulated datasets using the *mgwr* and *spreg* python packages, respectively. The default setting for MGWR is used with an adaptive bi-square kernel. For the SEM model, we adopted two approaches to specify the spatial weights matrix: (i) a queen contiguity-based (SEM Queen) and (ii) an *AIC*-based model selection procedure to select the number of nearest neighbors (SEM *AIC*-KNN). In order to calibrate an MLM model, a second-level framework is needed, and we designed two aggregated levels, one consisting of a 4 × 4 matrix with each cell containing 100 individuals, and the other consisting of an 8 × 8 matrix with each cell containing 25 individuals, as depicted in Figure 8.3. The MLMs are calibrated using the *lme4* R package (Bates et al., 2015).

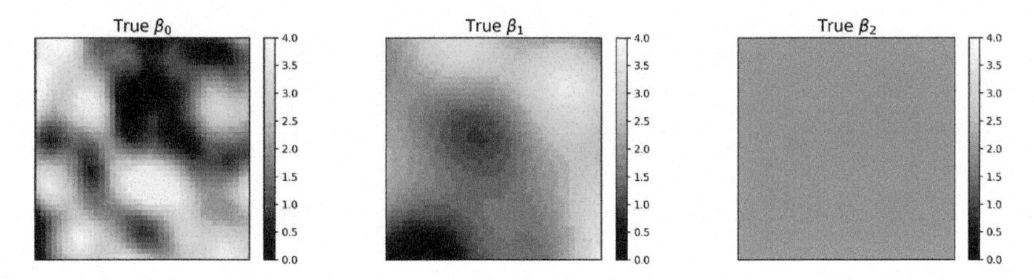

FIGURE 8.2
Three True Data Generating Processes Used in the Monte Carlo Simulation.

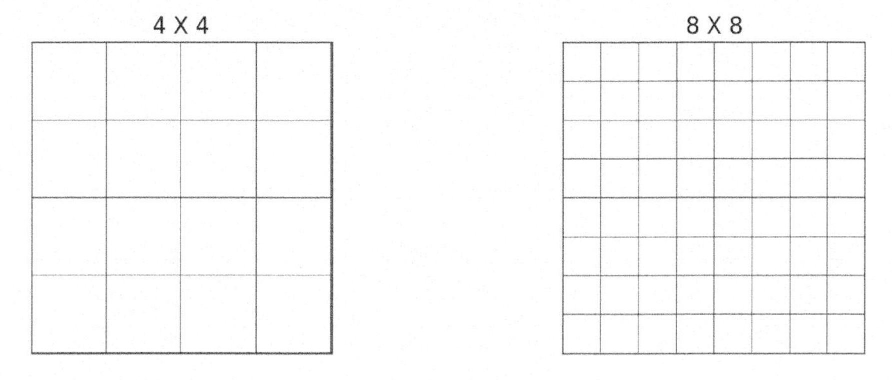

FIGURE 8.3
Two Second-Level Units Used in the Multilevel Model.

8.3.2 Comparison of MGWR and SEM

In order to compare the MGWR model with the SEM, the parameter estimates from both models were averaged across the 1,000 realizations of the Monte Carlo simulations and compared against the true values as shown in Figure 8.4. Compared to the true data generating processes, MGWR produces estimates of all three parameters that are highly accurate. The results of the two SEM models calibration show that although the spatially autoregressive error component shows a similar pattern to the local intercept, β_0, it cannot pick up any behavioral contextual effects in β_1.

To explore these results further, we calculated the amount of bias in the parameter estimates from both MGWR and SEM, and these are shown in Table 8.1. For each process, we computed two statistics, the mean, $E\left(\hat{\beta}-\beta\right)$, and the standard

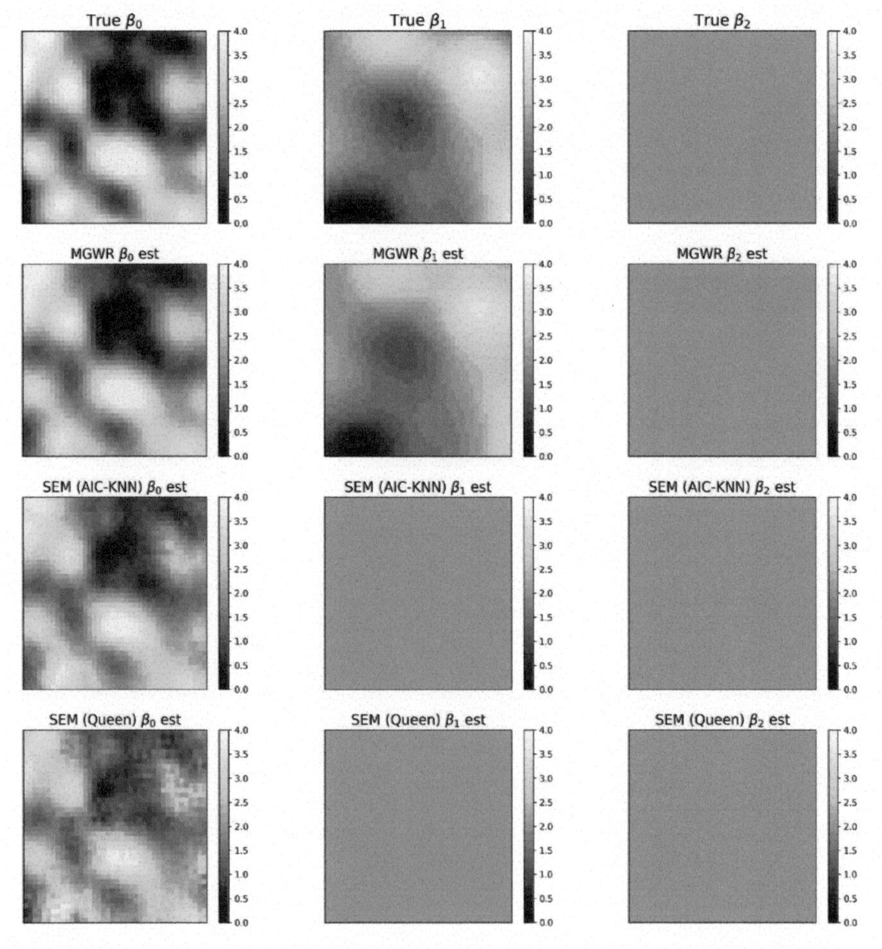

FIGURE 8.4
Averaged Parameter Estimates From MGWR and the SEM From the Monte Carlo Simulations.

TABLE 8.1

Mean and Standard Deviation of the Local Parameter Estimation Biases From MGWR and Two Versions of an SEM. The Last Row Depicts Moran's *I* of the Residuals.

	MGWR	SEM (*AIC*-KNN)	SEM (Queen)
$E(\hat{\beta}_0 - \beta_0)$	0.00	0.01	0.02
$\sigma(\hat{\beta}_0 - \beta_0)$	0.18	0.31	0.41
$E(\hat{\beta}_1 - \beta_1)$	0.02	0.03	0.05
$\sigma(\hat{\beta}_1 - \beta_1)$	0.12	0.77	0.77
$E(\hat{\beta}_2 - \beta_2)$	0.00	−0.02	−0.02
$\sigma(\hat{\beta}_2 - \beta_2)$	0.00	0.00	0.00
Moran's *I*	−0.05	−0.01	−0.05

deviation, $\sigma(\hat{\beta} - \beta)$, of the bias in the estimates across all the locations. The mean bias measures the accuracy of the estimate of the global trend in the process, while the standard deviation measures how accurately the remaining spatial pattern is captured. From Table 8.1, MGWR has smaller global trend estimation biases for all three surfaces compared to SEM. Regarding the spatial pattern estimation, for the local intercept, β_0, MGWR has a lower bias than the SEM (*AIC*-KNN) model (0.18 vs. 0.31). For the regional process β_1 that is associated with covariate x_1, MGWR has a much lower bias than the SEM (*AIC*-KNN) model (0.12 vs. 0.77), which is expected because SEM only estimates the global trend in β_1, ignoring its spatial heterogeneity. For the global process β_2, MGWR has minimal or no bias in estimating the global relationship, and SEM also has a very small bias. The spatial autocorrelation of the residuals, measured by Moran's *I*, was calculated, and the values for both models are very low, indicating that no spatial autocorrelation was left in the residuals. It is also worth noting that the use of a queen-based spatial weight matrix in SEM (Queen) increases the estimation bias of the spatial autoregressive component when comparing against the SEM (*AIC*-KNN) model. Therefore, it is more advisable to use a model selection approach such as the *AIC*-KNN approach to optimally identify the correct scale of the spatial error.

8.3.3 Comparison of MGWR and MLM

MGWR is compared to a multilevel model operating with two different second-level unit schemes as shown in Figure 8.3. Again, we averaged both the fixed and random effects for the multilevel models across the 1,000 realizations in the Monte Carlo simulation, and these are shown in Figure 8.5. It is clear that for the spatially varying processes, β_0 and β_1, multilevel models can approximate the spatial heterogeneity operating at the aggregated level but only crudely, and this is limited

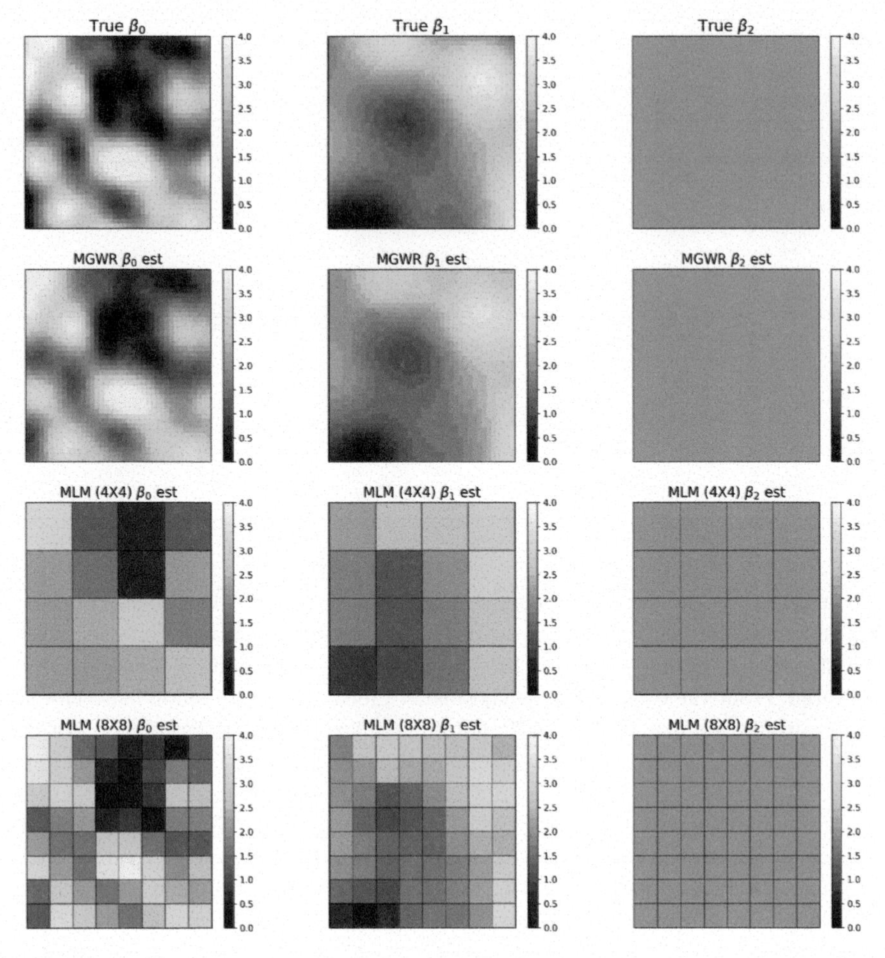

FIGURE 8.5
True and Averaged Parameter Estimates From MGWR and Multilevel Models in the Monte Carlo Simulation.

by the definition of the upper-level geographic divisions. Obviously, the finer the divisions available at the upper level, the better will be the representation of the spatial varying process, but this is a clear limitation of the MLM framework. For the global process, β_2, both multilevel model applications accurately estimate the global constant.

Both the mean and the standard deviation of the estimation bias of the multilevel models are shown in Table 8.2. For all three processes, the multilevel models accurately estimate the global trend in the processes, as indicated by the low mean bias ranging from −0.02 to 0.02. For the spatial heterogeneity in processes β_0 and β_1, using a finer resolution 8 × 8 grid for the second level, as expected, generates a smaller bias than using the 4 × 4 grid (β_0: 0.47 vs. 0.80; β_1: 0.23 vs. 0.37), especially

TABLE 8.2

Mean and Standard Deviation of the Local Parameter Estimation Bias for MGWR and Two Versions of MLM. The Last Row Depicts the Moran's *I* of the Residuals.

	MGWR	MLM (4 × 4)	MLM (8 × 8)
$E(\hat{\beta}_0 - \beta_0)$	0.00	−0.01	0.00
$\sigma(\hat{\beta}_0 - \beta_0)$	0.18	0.80	0.47
$E(\hat{\beta}_1 - \beta_1)$	0.02	−0.02	0.02
$\sigma(\hat{\beta}_1 - \beta_1)$	0.12	0.37	0.23
$E(\hat{\beta}_2 - \beta_2)$	0.00	−0.02	0.02
$\sigma(\hat{\beta}_2 - \beta_2)$	0.00	0.01	0.02
Moran's *I*	−0.05	0.30	0.07

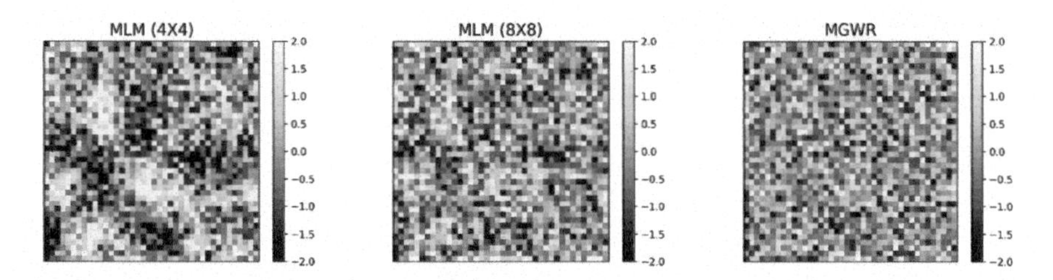

FIGURE 8.6
Residuals From Two Multilevel Models and MGWR.

for the more localized process β_0. Clearly, the finer the resolution of the second-level units, the less will be the bias in the local estimates, but it is unlikely that such a fine scale set of units would be available as to make the MLM results as useful as those of MGWR. Analysis of the Moran's *I* values indicates that spatial dependency is not removed from the residuals for the MLM with a 4 × 4 second-level specification. Figure 8.6 shows the residuals obtained from one realization of the Monte Carlo simulation, where the residuals from the multilevel model with 4 × 4 second-level units exhibit a much stronger autocorrelation than the residuals from the model calibrated with 8 × 8 second-level units.

8.4 Comparisons Based on Empirical Data

In this section, we applied MGWR, MLM, and SEM to the election dataset described in Chapter 7. For the MLM model, we fitted a varying intercept and slope model that allows for random effects to operate at the state level. Figure 8.7 shows a comparison

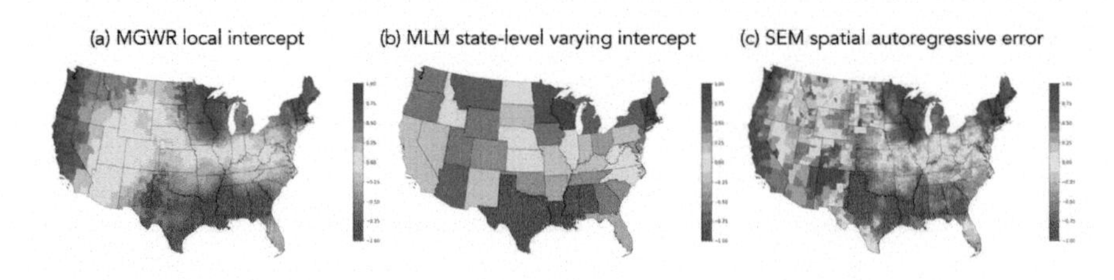

FIGURE 8.7
Estimates for the MGWR Local Intercept, the LML State-Level Varying Intercept and the SEM Spatial Autoregressive Error.

between the local intercept for MGWR, the state-level random intercept for MLM, and the spatial autoregressive error for SEM. We find strong similarities between the three estimates of the intrinsic contextual effects; i.e., contextual effects in counties in the southern states lead to a reduced vote for the Democratic Party whereas contextual effects in counties in the Pacific West, Upper Midwest, and Northeast lead to an increased vote for the Democratic Party. Figure 8.8 compares the behavioral contextual effects associated with the covariates. As SEM only produces global estimates, here we excluded it from the comparison and focus on MGWR and MLM. Again, the spatial heterogeneity appears to be similar though MLM operates at the state level while MGWR has county-level estimates. Disparities between the two sets of results from the crudeness of using states as the lowest level of the hierarchy in the MLM and the classic MLM model used here do not consider the spatial structure of the second-level unit. It is worth noting that spatial structure and associated effects can be introduced into MLM models, as in the work of Dong et al. (2015) and Wolf et al. (2021).

8.5 Comparison of MGWR and a Machine Learning Model

Machine learning, and more generally artificial intelligence, has been increasingly used to model spatial data, with many applications found in both social and environmental research (e.g., Jordan & Mitchell, 2015; Reichstein et al., 2019; Grimmer et al., 2021). While machine learning models have good predictive power, they are often considered black boxes and do not offer much interpretability. To address this issue, there have been recent developments in the field of explainable artificial intelligence (XAI) that aim to peek into the black box and understand the processes and mechanisms that lead to the model predictions. XAI methods can be broadly divided into global and local interpretation methods (Molnar, 2020). Global interpretation tends to explain the average AI behavior of all data, while local interpretation can look at the behavior of the model for each data point. For geographical studies, local interpretation methods provide an opportunity to visualize the spatial effects estimated by machine learning models, as individual

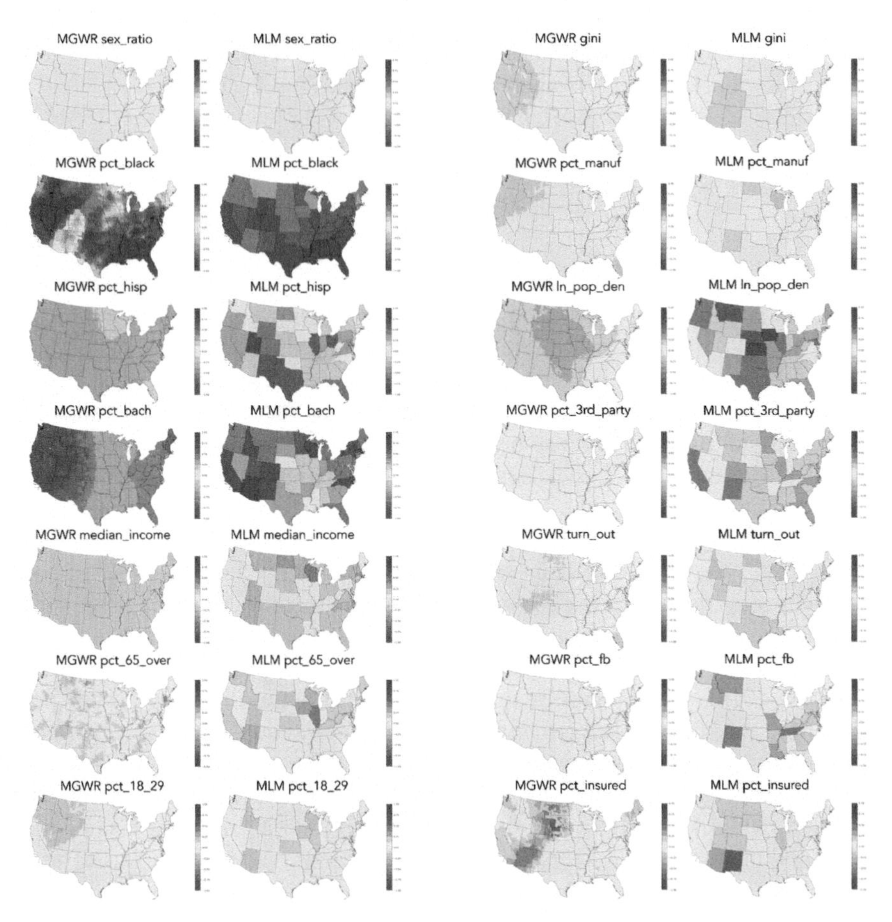

FIGURE 8.8
Comparison Between the MGWR Local Parameter Estimates and the MLM State-Level Varying Effects.

observations are georeferenced (Li, 2022). Currently, the most widely used local interpretation method is based on the Shapley value, originally developed in cooperative game theory, which fairly distributes the contributions of the players in a game (Shapley, 1953). In the context of machine learning, the Shapley value measures the marginal contribution of a covariate to a model prediction (Lundberg & Lee, 2017; Štrumbelj & Kononenko, 2014). The Shapley value for a covariate $\boldsymbol{X}_j$ in a model is given by

$$Shapley\left(\boldsymbol{X}_j\right) = \sum_{S \subseteq K \setminus \{X_j\}} \frac{s!(n-s-1)!}{k!}\left(f\left(S \cup \{j\}\right) - f(S)\right) \tag{8.10}$$

where k is the total number of covariates, $K \setminus \{\boldsymbol{X}_j\}$ is the set of all possible combinations of covariates excluding $\boldsymbol{X}_j$, $\boldsymbol{S}$ is a covariate set in $K \setminus \{\boldsymbol{X}_j\}$ with size s,

$f(\boldsymbol{S})$ is the model prediction using the covariates in $\boldsymbol{S}$, and $f(\boldsymbol{S}\cup\{j\})$ is the model prediction using the covariates in $\boldsymbol{S}$ plus covariate $\boldsymbol{X}_j$. The interpretation of the Shapely value is the average contribution of the covariate to the model prediction across all possible combinations of covariates. The resulting Shapley value provides an additive explanation to each model prediction:

$$\hat{y}_i = shap_0 + shap(X_{1i}) + shap(X_{2i}) + \ldots + shap(X_{ki}) \quad (8.11)$$

where $\hat{y}_i$ is the model prediction value for the observation i, $shap_0 = E(\hat{y})$ is the mean prediction across all observations, and $shap(X_{ji})$ refers to the Shapley value of the jth covariate for observation i, which represents the marginal contribution of the covariate to the prediction. In this way, the sum of all the Shapley values is equal to the difference between actual prediction and the average prediction. Such an additive specification has a similar interpretation to a linear regression model.

In this section, we used XGBoost (eXtreme Gradient Boosting) as an example of a machine learning model to replicate the contextual effects in the data generation process described in Section 8.3 and compared it to MGWR. XGBoost is a gradient boosting method that sequentially ensembles decision trees to minimize the model error using a gradient descent optimization algorithm (Chen & Guestrin, 2016). It is shown in the literature that a properly tuned XGBoost typically outperforms alternative methods such as random forest or deep neural network for supervised problems. We included the coordinates of each grid cell as additional two covariates in the model to account for spatial effect. The XGBoost model was fitted using the *python* package *xgboost* (Chen & Guestrin, 2016), and the hyperparameters were tuned using a Bayesian optimization package, *hyperopt* (Bergstra et al., 2013) nested within a five-fold cross-validation. Then Shapley values are calculated for the XGBoost model using the *python* package *shap*.

Shapley value-based coefficients were computed following Li (2022) by extracting the covariate contribution to model prediction from coordinates and coordinate interactions, and they are shown in Figure 8.9. When comparing the spatial pattern of parameter estimates, we find that the spatial variation can be approximated by XGBoost, but the estimated effect is not as smooth as that from MGWR. This is because XGBoost, as a tree-based model, does not have smooth decision boundaries, making the estimation more jagged and boxy. This is also reflected in the estimation bias shown in Table 8.3, where, although XGBoost has little bias in modeling the global effect, it has large biases in replicating the spatial variation in the effect. Since many spatial phenomena are continuous over space, this may be a potential drawback of adopting standard tree-based machine learning for modeling spatial processes. Special considerations for capturing spatial effects are needed in the future development of GeoAI (geospatial artificial intelligence) models (Janowicz et al., 2020; Li, 2020).

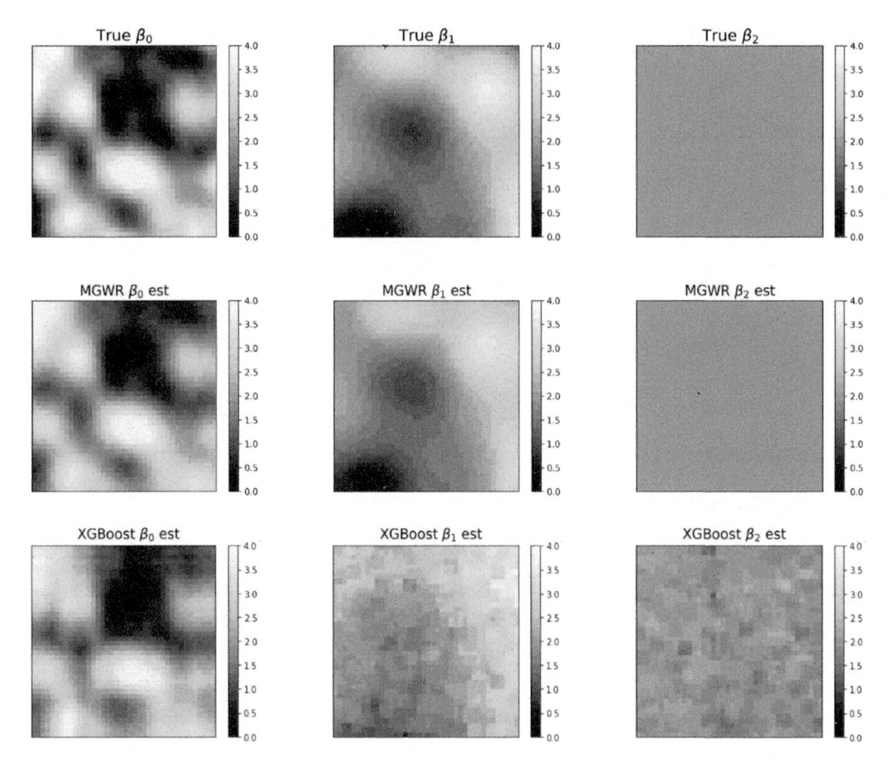

FIGURE 8.9
True Data Generating Processes, Averaged Parameter Estimates From MGWR, and Shapley-Based Parameter Estimates From XGBoost Models.

TABLE 8.3

Comparison of the Mean and Standard Deviation of the Parameter Estimation Biases Between MGWR and XGBoost.

	MGWR	XGBoost
$E\left(\widehat{\beta}_0 - \beta_0\right)$	0.00	0.07
$\sigma\left(\widehat{\beta}_0 - \beta_0\right)$	0.18	0.40
$E\left(\widehat{\beta}_1 - \beta_1\right)$	0.02	−0.01
$\sigma\left(\widehat{\beta}_1 - \beta_1\right)$	0.12	0.37
$E\left(\widehat{\beta}_2 - \beta_2\right)$	0.00	−0.01
$\sigma\left(\widehat{\beta}_2 - \beta_2\right)$	0.00	0.31
Moran's I	−0.05	−0.02

TABLE 8.4

Comparison of Model Approaches Incorporating Contextual Effects.

	Intrinsic	Behavioral
MGWR	Local intercept (β_0)	Local slopes $(\beta_{\neq 0})$
SEM	Spatial autoregressive error + global intercept $(\beta_0 + \lambda Wu)$	N/A
MLM	Global intercept + group-level varying intercept $(\beta_0 + \mu_0)$	Global slopes + group-level varying slopes $(\beta_{\neq 0} + \mu_{\neq 0})$
ML	Contribution of coordinates in the model	Interaction effects between coordinates and covariates

8.6 Summary

In this chapter, we compare MGWR with other spatial models that can capture contextual effects. Using Monte Carlo simulations and an empirical dataset on voting in the US Presidential election, we show that MGWR and other alternative modeling frameworks produce similar estimated spatial patterns of intrinsic contextual effects, with MGWR having the best accuracy. Table 8.4 summarizes different modeling approaches to incorporate intrinsic and behavioral contextual effects. MGWR estimates spatially varying local intercept and slopes, which measure intrinsic and behavioral contextual effects, respectively. SEM uses a spatial autoregressive error to represent intrinsic contextual effects but being a global model, it cannot capture any behavioral contextual effects. MLM, with its estimates of varying intercepts and slopes, is able to capture both intrinsic and behavioral contextual effects, but these are only allowed to vary at a predefined aggregated level. We show that MLM has large bias if the process operates at a more local scale than the aggregated level. Furthermore, using local interpretation methods, a machine learning model (XGBoost in our example) can approximate spatial effects in the data generating process through coordinate interactions, but the model is less accurate than MGWR. In summary, if the goal is to measure the individual-level spatial contextual effects, MGWR would appear to represent the best practice.

References

Anselin, L. (1988). *Spatial econometrics: Methods and models* (Vol. 4). Berlin: Springer Science & Business Media.

Anselin, L., & Bera, A. K. (1998). Spatial dependence in linear regression models with an introduction to spatial econometrics. In A. Ullah & D. E. A. Giles (Eds.), *Handbook of applied economic statistics* (pp. 237–290). Boca Raton, FL: CRC Press.

Bates, D., Mächler, M., Bolker, B., & Walker, S. (2015). Fitting linear mixed-effects models using lme4. *Journal of Statistical Software*, *67*(1), 1–48.

Bergstra, J., Yamins, D., & Cox, D. (2013). Making a science of model search: Hyperparameter optimization in hundreds of dimensions for vision architectures. In S. Dasgupta (Ed.), *ICML proceedings of the 30th international conference on machine learning* (pp. 115–123). Atlanta: JMRL.org.

Chen, F., Leung, Y., Mei, C.-L., & Fung, T. (2022). Backfitting estimation for geographically weighted regression models with spatial autocorrelation in the response. *Geographical Analysis*, *54*(2), 357–381.

Chen, T., & Guestrin, C. (2016). Xgboost: A scalable tree boosting system. In *Proceedings of the 22nd ACM SIGKDD International Conference on Knowledge Discovery and Data Mining* (pp. 785–794). New York: ACM.

Chi, G., & Zhu, J. (2019). *Spatial regression models for the social sciences*. Newbury Park, CA: SAGE.

Dong, G., Harris, R., Jones, K., & Yu, J. (2015). Multilevel modelling with spatial interaction effects with application to an emerging land market in Beijing, China. *PLoS One*, *10*(6), e0130761.

Duncan, C., Jones, K., & Moon, G. (1998). Context, composition and heterogeneity: Using multilevel models in health research. *Social Science & Medicine*, *46*(1), 97–117.

Grimmer, J., Roberts, M. E., & Stewart, B. M. (2021). Machine learning for social science: An agnostic approach. *Annual Review of Political Science*, *24*, 395–419.

Janowicz, K., Gao, S., McKenzie, G., Hu, Y., & Bhaduri, B. (2020). GeoAI: Spatially explicit artificial intelligence techniques for geographic knowledge discovery and beyond. *International Journal of Geographical Information Science*, *34*(4), 625–636.

Jordan, M. I., & Mitchell, T. M. (2015). Machine learning: Trends, perspectives, and prospects. *Science*, *349*(6245), 255–260.

LeSage, J., & Pace, R. K. (2009). *Introduction to spatial econometrics*. Boca Raton, FL: Chapman and Hall/CRC.

Li, W. (2020). GeoAI: Where machine learning and big data converge in GIScience. *Journal of Spatial Information Science*, *20*, 71–77.

Li, Z. (2022). Extracting spatial effects from machine learning model using local interpretation method: An example of SHAP and XGBoost. *Computers, Environment and Urban Systems*, *96*, 101845.

Lundberg, S. M., & Lee, S. I. (2017). A unified approach to interpreting model predictions. *Advances in Neural Information Processing Systems*, *30*, 4765–4774.

Ma, J., Li, C., Kwan, M. P., & Chai, Y. (2018). A multilevel analysis of perceived noise pollution, geographic contexts and mental health in Beijing. *International Journal of Environmental Research and Public Health*, *15*(7), 1479.

Molnar, C. (2020). *Interpretable machine learning*. Morrisville: Lulu.

Müller, S., Schüler, L., Zech, A., & Heße, F. (2022). GSTools v1.3: A toolbox for geostatistical modelling in python. *Geoscientific Model Development*, *15*(7), 3161–3182.

Orford, S. (2000). Modelling spatial structures in local housing market dynamics: A multilevel perspective. *Urban Studies*, *37*(9), 1643–1671.

Park, D. K., Gelman, A., & Bafumi, J. (2004). Bayesian multilevel estimation with poststratification: State-level estimates from national polls. *Political Analysis*, *12*(4), 375–385.

Reichstein, M., Camps-Valls, G., Stevens, B., Jung, M., Denzler, J., & Carvalhais, N. (2019). Deep learning and process understanding for data-driven Earth system science. *Nature*, *566*(7743), 195–204.

Shapley, L. S. (1953). 17. A value for n-person games. In Harold William Kuhn and Albert William Tucker (Eds.), *Contributions to the theory of games (AM-28), Volume II* (pp. 307–318). Princeton, NJ: Princeton University Press. https://www.degruyter.com/document/doi/10.1515/9781400881970/html

Štrumbelj, E., & Kononenko, I. (2014). Explaining prediction models and individual predictions with feature contributions. *Knowledge and Information Systems*, *41*(3), 647–665.

Twigg, L., Moon, G., & Jones, K. (2000). Predicting small-area health-related behaviour: A comparison of smoking and drinking indicators. *Social Science and Medicine*, *50*(7–8), 1109–1120.

Wolf, L. J., Anselin, L., Arribas-Bel, D., & Mobley, L. R. (2021). On spatial and platial dependence: Examining shrinkage in spatially dependent multilevel models. *Annals of the American Association of Geographers*, *111*(6), 1679–1691.

Yu, H., & Fotheringham, A. S. (2022). A multiscale measure of spatial dependence based on a discrete Fourier transform. *International Journal of Geographical Information Science*, *36*(5), 849–872.

Zahnd, W. E., & McLafferty, S. L. (2017). Contextual effects and cancer outcomes in the United States: A systematic review of characteristics in multilevel analyses. *Annals of Epidemiology*, *27*(11), 739–748.

9

Epilogue

9.1 Synopsis

The central tenet of this book is that under certain circumstances traditional global regression formulations are not appropriate to model the processes that lead to spatially varying outcomes. These circumstances are when the processes being modeled are not constant, as is assumed in a global model, but vary across space due to unmeasured, and possibly unmeasurable, contextual effects. The argument is made that global models potentially suffer from three problems when they are applied to processes, which exhibit spatial heterogeneity:

1. They will be misspecified by not accounting for unmeasurable contextual variables. This will create misspecification bias in any parameter estimate associated with a covariate that has a nonzero correlation with any of the omitted contextual effects.
2. They will be misspecified because the parameters of the model are fixed, yet in reality they vary across locations.
3. Unless residual dependency is explicitly modeled, they will almost certainly have residuals which are positively spatially autocorrelated, a property which will invalidate the regression assumption that the residuals are independent and which may lead to incorrect inferences being drawn from the calibrated model.

Local models such as MGWR overcome all three problems by allowing the parameters of the model to vary across locations. We describe in Chapter 1 how contextual effects are of two types, which are captured by separate elements of local models. *Intrinsic* contextual effects are those which relate to unmeasured attributes of a location that affect the dependent variable. In global models, these omitted effects are captured by the error term, leading to spatially dependent errors. In local models they are captured by the locally varying intercept with the error terms being independent. *Behavioral* contextual effects are those that lead to the marginal effect of a change in a covariate on y varying across locations. These contextual effects are captured in local models by spatially varying slope parameters. Both intrinsic and behavioral contextual effects can arise from a variety of sources, including traditions, persistent

DOI: 10.1201/9781003435464-9

adverse or beneficial conditions, customs, lifestyles, and psychological profiles common to an area that influence social norms, which in turn affect individual behavior, local media, the influence of family, friends and local organizations, and environmental conditions, all of which are reinforced by selective migration.

There are alternative modeling frameworks, such as spatial regime models, spatial error models, multilevel models, and machine learning models, that aim to capture contextual effects. However, these alternative frameworks are shown to be less flexible than MGWR and suffer from at least one of the following problems:

1. They cannot capture behavioral contextual effects;
2. The spatial variation in contextual effects has to be defined *a priori*;
3. They do not capture intrinsic contextual effects, including their uncertainty, as accurately as MGWR; and
4. They do not provide an intuitive measure of the spatial scale over which contextual effects vary.

MGWR achieves the latter through estimates of covariate-specific bandwidths, each of which is determined by a trade-off between the amount of bias in the local parameter estimates and their uncertainty (or variance). The optimal bandwidth denotes the limit where the data borrowed for the local regression reduce parameter uncertainty more than they increase parameter bias. Beyond the bandwidth, the addition of further data to the regression increases bias more than it decreases uncertainty. Hence the bandwidth is a measure of a property of spatially varying processes, which can be compared both between different processes and across applications to inform on the relative spatially varying nature of local processes. Increasingly large bandwidths indicate processes that are more stable over space: a global process in theory would have an infinitely large bandwidth.

Another way of viewing the bandwidth is to think of data being composed of '*information*' and '*misinformation*' in terms of the processes operating at location *i*. Data close to *i*, and which presumably are the product of processes similar to those being estimated at *i*, will contain more information than misinformation about the processes at *i* and so are included in the local regression. Data far from *i*, and which, if the processes exhibit spatial heterogeneity, are likely to be the product of very different processes to those at *i*, will contain more misinformation than information and so are excluded from the location regression at *i*. The bandwidth is thus the point in space from *i* beyond which data contain more misinformation than information about the processes at *i*.

Both these conceptualizations of the bandwidth suggest it can be used as an indicator of the spatial scale over which a process is relatively stable. In addition, two long-standing scale issues, the modifiable areal unit problem and Simpson's paradox, can be understood through the lens of local models and spatial processes. Importantly, if processes are spatially heterogeneous, then different combinations of processes will be captured when data are aggregated to different spatial scales. This insight is particularly crucial for understanding the issue of reproducibility and replicability in spatial analysis. Rather than invalidating the conclusions reached from the analysis of spatial relationships, the MAUP and Simpson's paradox instead reflect that when we model at different scales, we are answering different questions.

Given the estimation of a set of covariate-specific bandwidths, we show how the calibration of a model by MGWR produces local parameter estimates on which inference can be performed allowing for both multiple hypothesis testing and dependency between the hypotheses. A calibration checklist is provided for defensible empirical applications of MGWR, and an example based on voting in the 2022 US Presidential election is presented.

9.2 Extensions

The concept of data borrowing used in GWR and MGWR can be extended to other forms of spatial statistical analysis. There are examples in the literature, for instance, of GW principal components analysis (Harris et al., 2011), GW discriminant analysis (Brunsdon et al., 2007), GW summary statistics (Brunsdon et al., 2002), and GW spatial interaction models (Kordi & Fotheringham, 2016). In each application, the concept is the same—instead of obtaining one set of what are essentially spatially averaged results, why not obtain results, which are specific to each location by borrowing data from nearby locations and weighting these between 0 and 1 according to how far away they are from the focal point of the analysis. Of course, in these extensions to the basic MGWR framework, if the weighting function cannot be optimized in some way, the results will be of limited use. Several extensions are described in the next section that do derive an optimized bandwidth for the weighting function and directly build upon the MGWR framework.

9.2.1 Multiscale Geographically Weighted Generalized Linear Modeling (MGWGLM)

A basic GWR or MGWR routine assumes that the conditioned response (error term) is normally distributed (i.e., Gaussian), which is useful when modeling continuous phenomena with certain characteristics. However, it may be more reasonable to assume that the conditional response follows an alternative distribution when the assumptions of normality are not met or when the phenomenon being modeled is not continuous. The generalized linear modeling (GLM) framework permits additional types of conditional responses from among the family of exponential probability distributions by using different link functions to transform the response variable during model calibration. The GLM framework has been incorporated into the GWR framework to support non-Gaussian models, and this extension was initially demonstrated for geographically weighted Poisson regression to model counts but has also been adopted for geographically weighted logistic regression to model binary (i.e., binomial) outcomes, as well as geographically weighted beta regression, geographically weighted negative binomial regression, and others (Fotheringham et al., 2002; Nakaya et al., 2005; da Silva & Rodrigues, 2014; da Silva & de Oliveira Lima, 2017).

It is also possible to incorporate the GLM framework into the MGWR framework to support non-Gaussian MGWR models. However, since MGWR uses the generalized additive modeling framework (GAM) and a backfitting calibration procedure, special care needs to be taken when combining aspects of the algorithms from the GLM, GAM, and GWR routines. Successfully combining these three frameworks is an area of important future research that would make it possible to carry out MGWR model calibrations for counts and binary outcomes in addition to continuous variables. For a start in this direction, see Sachdeva et al. (2023) who have developed a Poisson version of MGWR, which they apply to counts of COVID cases.

9.2.2 Multiscale Geographically and Temporally Weighted Regression (MGTWR)

The approach of borrowing data from "nearby" locations can also be extended to the temporal domain to explore heterogeneity in processes across time in addition to space. In this case, data also contain information about when an event occurred or when the information was collected and separation can also be conceptualized in terms of the amount of time that passes between observations. Those observations that occurred recently compared to the observation at the calibration location are considered more influential in the regression than those that occurred in the more distant past. Several methodologies have been developed for geographically and temporally weighted regression (GTWR) to simultaneously consider both the geographical and temporal domains. An initial methodology proposed a three-dimensional weighting function (two for Euclidean geometric distances and one for time) in order to create a spatiotemporal kernel (Huang et al., 2010). However, space and time are typically measured in different units and have their own separate scales, which make it difficult to interpret the associated bandwidth and disentangle the role of either space or time within the model. A subsequent methodology adopted an alternative approach of calibrating and applying a geographic weighting function with different bandwidths (i.e., decay of influence) for data from different time intervals away from the calibration observation (Fotheringham et al., 2015). This allows the bandwidth to maintain its original interpretation from GWR, as well as becoming flexible enough to consider the influence of time on local regressions. As a result, the bandwidths from this methodology generally become smaller for larger time intervals, which means that observations become less related the further away they are in time or space, but that two points separated by the same distance will have less of an influence on each other when they are separated by longer time intervals than by shorter time intervals (Fotheringham et al., 2015). GTWR is most often applied in the context of house price modeling but has also been applied in other disciplines such as epidemiology and transportation modeling (Chen et al., 2021; Huang & Xu, 2021; Hong et al., 2021).

GTWR shares the same limitation as GWR regarding the assumption that the spatial heterogeneity of processes occurs at a single scale. This assumption is perhaps even more tenuous in the spatiotemporal domain because it becomes increasingly unlikely that all processes occur at both the same spatial scale and the same temporal scale. As alluded to earlier, by incorporating both space and time, which occur for different types of units at different scales, there is already an implicit notion of multiple scales being incorporated into the analysis. Extending MGWR into the temporal domain to create MGTWR allows for the multiscale notion to become explicit and for the relaxation of the assumption that the same spatial and temporal scales must apply to each relationship in the model (Wu et al., 2019; Zhang et al., 2021; Liu et al., 2021). In addition to having some relationships that occur locally and some that occur globally, it also becomes possible to have some relationships that occur over shorter time horizons and others that occur over longer time horizons.

9.2.3 Spatially Weighted Interaction Models

Another extension of the GWR framework focuses on the local modeling of spatial interaction phenomena or the aggregate flows of people, goods, information, or value (Kordi & Fotheringham, 2016). Since spatial interaction models have their own rich history, this extension is aptly referred to as spatially weighted interaction models (SWIM). Although spatial interaction models can be calibrated using regression, it is not possible to directly apply the GWR concepts and algorithms to flow data because the weighting function in GWR considers observation locations embedded in two-dimensional space while spatial interaction models consider observations that occur between locations and are embedded in four-dimensional space. Two solutions have been proposed to overcome this. The first solution seeks to simplify the dimensionality issue by only focusing the geographical weights on either the destination locations or the origin locations of the observations. In this way, it is possible to extend the concept of data borrowing and explore process nonstationary in some of the model terms but not all of them. Furthermore, spatial interaction models also typically include variables that explain the distance or cost to overcome the separation between each origin and destination, and the origin-focused or destination-focused strategies are not able to accommodate these factors because they cannot be simplified to two dimensions. Instead, the second solution frames the spatial weights in four dimensions so that data can be borrowed from neighboring observations adaptively based on both the origins and destinations of flow observations. This flow-focused strategy allows all variables to be geographically weighted. Since the advent of MGWR, efforts have already produced a multiscale flow-focused SWIM to analyze transportation flows (Zhang et al., 2019), though more work is needed on this recent modeling extension to better understand the role of multiple scales in spatial interaction modeling.

9.2.4 Local Bandwidths

One of the core ideas throughout this book is that when addressing issues related to human behavior it is reasonable to assume that the processes driving such behavior might vary across space and that the nature of this variation is best captured across multiple scales. The MGWR framework allows these diverse scales to be adaptively identified and combined by allowing each relationship in the model to be expressed at a potentially unique scale compared to other relationships. It is also possible to imagine that the spatial variability of a particular process might itself vary over space so that it varies over short distances in some locations and is relatively stable in others. In this case, the multiscale heterogeneity can be captured by allowing location-specific bandwidths that generate unique weighting functions depending upon where a process is being measured. This type of analysis could be achieved by simply dividing a study area up into regions of interest, carrying out a bandwidth selection routine for each region using only observations within that region as calibration locations, and then applying the regional bandwidths in a piecewise fashion to obtain the final model. However, this strategy assumes that we have *a priori* knowledge about how to divide the data into discrete regions and also suffers from edge effects due to the discrete and potentially arbitrary boundaries of the regions. A more flexible strategy involves generating location-specific bandwidths for each observation in the study area, which solves both of the above issues because then there would be no need to define regions and the bandwidths would be allowed to vary smoothly. Some initial work has been done in this direction within the GWR framework where the location-specific bandwidths are assumed to be the same for each relationship in the model (Páez et al., 2002; Comber et al., 2018), but none of these proposed extensions uses a global optimization criterion to select an optimal set of bandwidths or accommodates bandwidths that vary by both location and relationship. Overcoming these limitations would likely require novel algorithms for high dimensional optimization, as well as the incorporation of regularization techniques to ensure the model does not become too flexible—bandwidths varying by both location and relationship that become excessively local would imply the consumption of more degrees of freedom than are typically available. Developing an MGWR model with local bandwidths therefore remains an open challenge.

9.2.5 Machine Learning

Some parallels between the results from MGWR and from machine learning methods were elucidated in Chapter 8 when we considered various models that incorporated context, narrowing the gap between these two frameworks. Another similar trend that will likely further narrow the gap is the adoption of geographically weighted concepts to create spatially explicit machine learning methods. Contributions of this kind typically fall into two different categories. The first category involves a straightforward incorporation of geographically weighting as a data-borrowing technique to create local subsets of data and then separately run a standard machine learning algorithm on each subset. This strategy has been adopted to develop a geographically weighted random forest technique (Georganos et al., 2019). The second category involves embedding the geographic weighting function within machine learning algorithms and has resulted in both geographically weighted and geographically and temporally weighted artificial neural network techniques (Deng et al., 2017; Wu et al., 2021; Hagenauer & Helbich, 2022). Given the popularity and simplicity of geographic weighting, the increasing number of machine learning techniques, and the growing interest in explainable and interpretable machine learning and artificial intelligence, it seems likely that these trends will also continue to develop and expand to include multiscale variations.

9.3 Prediction with MGWR

Although GWR and MGWR models are typically calibrated and evaluated using each location in a sample where observations are available, it is also possible to predict the values of the dependent variable at a location where it is unmeasured when there are observations available on the independent variables. This entails calibrating a model for observations at locations where both the dependent variable and independent variables are measured. Then the optimal bandwidth(s) from that model can be used to create a data weighting scheme for use at the locations where the dependent variable is not measured. By borrowing data on both the dependent variable and independent variables, it is possible to generate parameter estimates at these unmeasured locations. Finally, observations on the independent variables can be combined with these parameter estimates in order to produce predictions for these locations that would otherwise not have information available about the dependent variable. This type of spatial out-of-sample prediction using GWR has been applied in many domains (Selby & Kockelman, 2013; Hu et al., 2013; Wang et al., 2014; Chen et al., 2018; Lin & Billa, 2021) and has been compared and combined with various geostatistical techniques and interpolation methods (Harris et al., 2010a, 2010b; Ye et al., 2017). However, only recently was it extended to incorporate multiple scales using the MGWR framework (Liu et al., 2021). Since MGWR allows covariate-specific bandwidths, this means that each explanatory variable can potentially influence the out-of-sample predictions to a different degree, and while initial results are promising, further research is necessary to demonstrate the advantages of using MGWR over other methods for prediction.

9.4 Implications

Local modeling represents a profound shift in spatial analytic reasoning and capability. Where the processes being modeled are spatially heterogeneous, the traditional "*one size fits all*" approach of global modeling is inappropriate, and local models will be more accurate, considerably so in many instances. The recognition that the marginal impact of a change in a covariate on y may be spatially varying has major implications for resource allocation and policy formulation. As in precision agriculture, why treat every location in the same way? The impact of changing a covariate, x, on y will have greater impact in some locations than in others, in effect bringing greater rewards for the same investment. If changing x has only a very minor impact on y in certain areas, why pursue the cost and effort of trying to change x in such locations? Allocating limited resources to the areas where they will be most effective is surely a universal administrative goal, and it is one where local modeling clearly can play a key role.

Beyond the obvious advantages of more efficient resource allocation and better understanding of spatial processes, the recognition that processes might not be the same everywhere has several other profound implications for spatial analysis.

9.4.1 Spatial Regression Models

Given that local models capture the role of context in generating dependency between data and processes, one obvious implication concerns the need for spatial regression models. The raison d'être for spatial regression modeling is to deal with the problem of spatially autocorrelated error terms, which is a problem frequently encountered in the global modeling of spatial data. However, there is considerable empirical evidence to suggest that any error dependency reported in a global model is greatly reduced to the point of statistical insignificance when a local version of the model is calibrated (Zhang et al., 2005, 2009; Osborne et al., 2007; Windle et al., 2010; Gao & Li, 2011; Sá et al., 2011; Miller, 2012). It is not difficult to understand why this happens—the local intercept in MGWR performs the same function as a locally weighted error function in a spatial error model. However, local modeling is to be preferred over spatial regression modeling for three reasons. One is that the local model would be expected to outperform a spatial error model in many applications because the former not only incorporates the influence of context but also allows the relationships being modeled to vary locally. A second is that in local models the spatial weighting matrix for the assessment of context dependency is derived from the data, whereas it typically has to be defined arbitrarily in the calibration of a spatial error model (although see spatial Gaussian copula regression [Masarotto & Varin, 2017] for an exception). The third is that a local model such as MGWR generates an optimal bandwidth parameter for the local estimates of the intercept, which can be interpreted as an indicator of the spatial scale over which context influences behavior (see Chapter 4).

9.4.2 The Modifiable Areal Unit Problem

Local models bring attention to the modeling of processes and specifically to the possibility that the unseen processes, which produce the data we observe, may vary over space. The notion that processes might vary over space has implications for a long-standing problem often encountered in the statistical analysis of spatial data—the modifiable areal unit problem (MAUP). The MAUP occurs when the same data are aggregated to different zoning systems and then these different zoning systems yield significantly different results when the same model is calibrated (Gehlke & Biehl, 1934; Openshaw, 1984; Fotheringham & Wong, 1991; Cressie, 1996). In most previous research on the MAUP, the issue has been treated as a data problem—that is, the aggregation of the data into zones alters some statistical property of the data, which in turn affects the results of models calibrated with the aggregated data (Gehlke & Biehl, 1934; Arbia, 1989; Amrhein, 1995; Dark & Bram, 2007). However, local models allow us to see this problem as process-related rather than data related. That is, the data we observe at the individual level result from processes at that scale that may vary over space. Different aggregations of these data will therefore lead to different combinations of processes at the aggregated level, which will yield different model calibration results. If processes are stable over space, the sensitivity of the results to MAUP will be relatively mild, whereas when processes become increasingly spatially nonstationary, the effect of the MAUP will intensify. Variations in the results of calibrating global regression models on different aggregations of the same underlying spatial data are therefore to be expected if the scale of the analysis changes and the processes that are being examined are spatially varying. This expectation of inconsistency in results is also critical to our understanding of reproducibility in spatial modeling of human behavior in that truly replicable research using aggregated data might be unattainable.

9.4.3 Simpson's Paradox

A more extreme version of the MAUP is encountered as Simpson's paradox where the signs of relationships vary with scale such that results from global and local models might lead to contradictory inferences (Simpson, 1951; Blyth, 1972; Alin, 2010). The paradox is usually reported in aspatial contexts, but it can also occur when the results of a global and local calibration of the same model are compared. It is possible, for example, to report a significant positive global parameter estimate for a conditioned relationship but a set of significant negative local parameter estimates for the same conditioned relationship. Less dramatic, but more usual, is the situation where a global parameter is not significant, yet many of the local estimates of the same parameter are significant or vice versa. The question is then: which set of results do we believe? Again, the problem is related to the scale of the analysis, and, in fact, both results can be correct because the analyses conducted at different scales answer different questions. For instance, consider household burglaries. At the level of individual housing units there may be an inverse conditioned relationship between the probability of a residence being burgled and it being vacant—why burgle an empty house? Whereas at the level of a neighborhood there might be a positive conditioned relationship between the burglary rates and vacant dwellings—fewer houses in the neighborhood have occupants to deter the burglary of occupied properties. Although Simpson's paradox has largely been a non-issue in spatial analysis, with the growing popularity of local modeling applications, and the inevitable comparisons between local and global model calibration results, it seems likely that more instances of Simpson's paradox will be encountered. As with the MAUP, the existence of the paradox is not a cause for concern; it merely reflects the fact that different questions are being asked when the same model is applied at different spatial scales.

9.4.4 Local Models and Replicability

There has been substantial discussion about replicability in the social and medical sciences prompted in large part by the fact that it is often difficult to reach consistent findings across different applications of the same model (*inter alia* Ioannidis, 2005; Camerer et al., 2016; Baker, 2016; Ioannidis et al., 2017).[1] Nowhere is this more true than in spatial analysis where the results of calibrating a model with data drawn from one area are often inconsistent with the results of calibrating the same model with data drawn from a different area (Kedron, 2021a, 2021b; Nüst et al., 2018). This has led, understandably, to a lack of belief in the robustness of empirically based social science research. However, the advent of local models and the increasing recognition that context can have a major impact on the determinants of human behavior suggest that it is our expectation of reproducibility that is at fault rather than the models we employ. Why should we expect model calibrations to yield similar results when those calibrations are based on data from different locations if contextual effects play an important role in our behavior? This also suggests that models calibrated with data in one location should not be applied in another location unless contextual effects are either absent or fully accounted for in the model; a new calibration should be undertaken. Clearly this is often time-consuming and costly, but the alternative is applying a model with parameter estimates that might yield highly misleading results. It would be rewarding if local models could be extrapolated to other areas but currently this remains elusive.

9.4.5 A Bridge between Idiographic and Nomothetic Approaches

Dating back to at least the Hartshorne-Schaefer debate, the geographic literature has long examined whether, and in what circumstances, we might expect geographic processes to remain stable, thereby allowing the study of those processes to be replicable (Hartshorne, 1939; Schaefer, 1953; Hartshorne, 1955). A schism arose, which persists to this day, with adherents of a "place-based", idiographic geography on one side and those who believed that regularities across space could be identified and measured through nomothetic approaches. For the former, space is seen as a major factor affecting people's behavior and hence trying to identify regularities in such behavior is futile. For the latter, space is viewed as a container in which regularities could be identified through quantitative modeling and spatial analytics. Both points of view have merit: it seems highly plausible that spatial context (i.e., "place") could have a major effect on behavior (see Chapter 1), but equally it would be useful to be able to quantify the determinants of behavior to enable normative prediction and understanding. Local models provide this very useful bridge between idiographic and normative approaches to understanding spatial behavior in that they allow the impacts of spatial context to be included in a nomothetic framework where regularities can be modeled. Indeed, the impact of spatial context, rather than left as a loose, rather nebulous, concept, can now be quantified and compared with the impact of other factors on behavior, as we show in Chapter 7 in the context of political voting. Further the effect of context is shown to be separable into two components in local models such as MGWR. Intrinsic contextual effects are those attributes of locations, which have a direct impact on y but which cannot be measured and hence are not included in a model. These are identified by the estimated location-specific intercepts. Behavioral contextual effects are those factors that impact the marginal change in y due to a change in covariate x. These are identified by significant variations in the location-specific slope parameter estimates. A great deal of research is now possible into the role of spatial context through the calibration of local models.

9.4.6 A Geography of Processes

For much of its history, spatial analytics has tended to focus on spatial data, either through mapping or by investigating properties such as spatial dependence or variability. Local modeling changes this perspective to one of spatial processes. The calibration of a local model produces sets of location-specific parameter estimates, standard errors, goodness-of-fit statistics, and other diagnostics, which can all be mapped to display facets of processes rather than data. Each set of local parameter estimates, for example, displays the conditioned relationship between y and a specific covariate x, which is a proxy for the unseen and often unseeable process that links these two sets of data. Consequently, the calibration of local models leads to a focus on processes, and through mapping and analyzing location-specific parameter estimates, we can get a sense of the spatial heterogeneity in such processes. We need to get used to seeing mapped values as not just the outcome of processes but also as indicators of the processes themselves.

9.5 The Future

Despite the enormous progress made in local spatial modeling over the past two decades, the field continues to evolve rapidly. For example, current research is focused on generalizing and extending the spatial weighting function, extending the range of models that can be made local, and extending the notion of 'local' to bandwidths in order to produce not only covariate-specific bandwidths but also location-specific, covariate-specific bandwidths. The latter is a natural evolution in the development of local spatial models, since if we recognize that processes might vary across space, is it possible that this variation itself might not be constant over space? Might processes vary at different spatial scales in different locations, being relatively constant in some regions but varying in others? Furthermore, the primary weighting function within the (M)GWR framework is spatial, based on the reasonable idea that data points closer in space are more likely to have been generated by more similar processes than data at locations further apart. However, other kinds of dependencies might exist such as in the temporal dimension, and expanding local modeling frameworks to include other types of weighting functions is potentially useful. For instance, expanding the GWR framework temporally has been shown to be more informative (Fotheringham et al., 2015; Huang et al., 2010). Further, a contextual weighting scheme, which weights observations based on similarities in the covariates used in the model in addition to spatial proximity, has been proposed by Harris et al. (2013). Finally, processes, especially those pertaining to social sciences, can also be expected to be dependent not just on space but on social networks formed on the internet, which could potentially affect human behavior similarly. There is potential therefore for developing local models that allow the inclusion of such nuances in terms of process dependency and spatial weighting and which pose novel statistical and computational challenges.

Another set of questions awaiting further research lies in the nature of '*what is context?*' Here we identify context as having two components, which we term 'intrinsic' and 'behavioral'. The former captures the illusive nature of attributes that are missing from the model. In some cases, we might suspect what these attributes are but cannot measure them; in other cases, we might have no idea what they are but recognize, particularly when dealing with the complexities of human behavior, we are unlikely to have captured everything that is pertinent in our model. A question in any application is: *how important are these missing attributes?* The latter relates to the marginal impacts on y of changes in a covariate x being spatially heterogeneous. Why, for example, might house prices be more sensitive to changes in the age of the property in some areas than in others, *ceteris paribus*, or why might preference for a political party be more sensitive to age in some areas? These are complex social questions that local modeling highlights. To what extent there exists some form of ground-truthing is another important research area—how can we support the evidence we obtain from local models in the form of spatially heterogeneous processes?

Finally, the last two decades have seen a revolution of sorts in terms of how spatial processes are perceived and modeled, which has opened new lines of inquiry regarding the nature of spatial processes. This inquiry has come from different directions and has led to different types of local models being formulated, but regardless of the statistical framework shaping these models, the goal is the same: to allow for, and to investigate, spatial variations in conditioned relationships, which provide evidence of spatially varying processes. However, much is still to be done, and hopefully this book will stimulate further research into the intriguing issue of how location affects behavior.

Note

1. Here we use the US National Science Foundation definition of replicability as being the ability of a researcher to duplicate the results of a prior study if the same procedures are followed but new data are collected (Bollen et al., 2015).

References

Alin, A. (2010). Simpson's paradox. *WIREs Computational Statistics*, *2*(2), 247–250.

Amrhein, C. G. (1995). Searching for the elusive aggregation effect: Evidence from statistical simulations. *Environment and Planning A: Economy and Space*, *27*(1), 105–119.

Arbia, G. (1989). *Spatial data configuration in statistical analysis of regional economic and related problems*. Berlin: Springer.

Baker, M. (2016). 1500 scientists lift the lid on reproducibility. *Nature*, *533*(7604), 452–454.

Blyth, C. R. (1972). On Simpson's paradox and the sure-thing principle. *Journal of the American Statistical Association*, *67*(338), 364–366.

Brunsdon, C., Fotheringham, A. S., & Charlton, M. E. (2002). Geographically weighted summary statistics—a framework for localised exploratory data analysis. *Computers, Environment and Urban Systems*, *26*, 501–524.

Brunsdon, C., Fotheringham, A. S., & Charlton, M. E. (2007). Geographically weighted discriminant analysis. *Geographical Analysis*, *39*(4), 376–396.

Camerer, C. F., Dreber, A., Forsell, E., Ho, T. H., Huber, J., Johannesson, M., Kirchler, M., Almenberg, J., Altmejd, A., Chan, T., & Heikensten, E. (2016). Evaluating replicability of laboratory experiments in economics. *Science*, *351*(6280), 1433–1436.

Chen, L., Ren, C., Zhang, B., Wang, Z., & Xi, Y. (2018). Estimation of forest above-ground biomass by geographically weighted regression and machine learning with Sentinel imagery. *Forests*, *9*(10), 582.

Chen, Y., Chen, M., Huang, B., Wu, C., & Shi, W. (2021). Modeling the spatiotemporal association between COVID-19 transmission and population mobility using geographically and temporally weighted regression. *GeoHealth*, *5*(5).

Comber, A., Wang, Y., Lu, Y., Zhang, X., & Harris, P. (2018). Hyper-local geographically weighted regression: Extending GWR through local model selection and local bandwidth optimization. *Journal of Spatial Information Science*, *17*, 63–84.

Cressie, N. (1996). Change of support and the modifiable areal unit problem. *Faculty of Informatics—Papers (Archive)*, 159–180.

da Silva, A. R., & de Oliveira Lima, A. (2017). Geographically weighted beta regression. *Spatial Statistics*, *21*, 279–303.

da Silva, A. R., & Rodrigues, T. C. V. (2014). Geographically weighted negative binomial regression—incorporating overdispersion. *Statistics and Computing*, *24*(5), 769–783.

Dark, S. J., & Bram, D. (2007). The modifiable areal unit problem (MAUP) in physical geography. *Progress in Physical Geography: Earth and Environment*, *31*(5), 471–479.

Deng, M., Yang, W., & Liu, Q. (2017). Geographically weighted extreme learning machine: A method for space—time prediction. *Geographical Analysis*, *49*(4), 433–450.

Fotheringham, A. S., Brunsdon, C., & Charlton, M. E. (2002). *Geographically weighted regression: The analysis of spatially varying relationships*. London: Wiley.

Fotheringham, A. S., Crespo, R., & Yao, J. (2015). Exploring, modelling and predicting spatiotemporal variations in house prices. *Annals of Regional Science*, *54*(2), 417–436.

Fotheringham, A. S., & Wong, D. W-S. (1991). The modifiable areal unit problem in multivariate statistical analysis. *Environment and Planning A*, *23*(7), 1025–1044.

Gao, J., & Li, S. (2011). Detecting spatially non-stationary and scale-dependent relationships between urban landscape fragmentation and related factors using geographically weighted regression. *Applied Geography*, *31*(1), 292–302.

Gehlke, C. E., & Biehl, K. (1934). Certain effects of grouping upon the size of the correlation coefficient in census tract material. *Journal of the American Statistical Association, 29*(185A), 169–170.

Georganos, S., Grippa, T., Niang Gadiaga, A., Linard, C., Lennert, M., Vanhuysse, S., Mboga, N., Wolff, E., & Kalogirou, S. (2019). Geographical random forests: A spatial extension of the random forest algorithm to address spatial heterogeneity in remote sensing and population modelling. *Geocarto International*, 1–16.

Hagenauer, J., & Helbich, M. (2022). A geographically weighted artificial neural network. *International Journal of Geographical Information Science*, *36*(2), 215–235.

Harris, P., Brunsdon, C., & Charlton, M. E. (2011). Geographically weighted principal components analysis. *International Journal of Geographical Information Science*, *25*(10), 1717–1736.

Harris, P., Charlton, M. E., & Fotheringham, A. S. (2010b). Moving window kriging with geographically weighted variograms. *Stochastic Environmental Research and Risk Assessment*, *24*(8), 1193–1209.

Harris, P., Fotheringham, A. S., Crespo, R., & Charlton, M. E. (2010a). The use of geographically weighted regression for spatial prediction: An evaluation of models using simulated data sets. *Mathematical Geosciences*, *42*(6), 657–680.

Harris, R., Dong, G., & Zhang, W. (2013). Using contextualized geographically weighted regression to model the spatial heterogeneity of land prices in Beijing, China. *Transactions in GIS*, *17*(6), 901–919.

Hartshorne, R. (1939). The nature of geography: A critical survey of current thought in the light of the past. *Annals of the Association of American Geographers*, *29*(3), 173–412.

Hartshorne, R. (1955). 'Exceptionalism in geography' re-examined. *Annals of the Association of American Geographers*, *45*(3), 205–244.

Hong, Z., Mei, C., Wang, H., & Du, W. (2021). Spatiotemporal effects of climate factors on childhood hand, foot, and mouth disease: A case study using mixed geographically and temporally weighted regression models. *International Journal of Geographical Information Science*, *35*(8), 1611–1633.

Hu, X., Waller, L. A., Al-Hamdan, M. Z., Crosson, W. L., Estes, M. G., Estes, S. M., Quattrochi, D. A., Sarnat, J. A., & Liu, Y. (2013). Estimating ground-level PM2.5 concentrations in the southeastern U.S. using geographically weighted regression. *Environmental Research*, *121*, 1–10.

Huang, B., Wu, B., & Barry, M. (2010). Geographically and temporally weighted regression for modeling spatio-temporal variation in house prices. *International Journal of Geographical Information Science*, *24*(3), 383–401.

Huang, Y., & Xu, W. (2021). Spatial and temporal heterogeneity of the impact of high-speed railway on urban economy: Empirical study of Chinese cities. *Journal of Transport Geography*, *91*, 102972.

Ioannidis, J. P. (2005). Why most published research findings are false. *PLoS Medicine*, *2*(8), e124.

Ioannidis, J. P., Stanley, T. D., & Doucouliagos, H. (2017). The power of bias in economics research. *The Economic Journal*, *127*(605), F236–F265.

Kedron, P., Frazier, A., Goodchild, M. F., Fotheringham, A. S., & Li, W. (2021a). Reproducible and replicable geospatial research: Where are we and where might we go? *International Journal of Geographic Information Science*, *35*(3), 427–445.

Kedron, P., Frazier, A., Trgovac, A., Nelson, T., & Fotheringham, A. S. (2021b). Reproducibility and replicability in geographical analysis. *Geographical Analysis*, *53*(1), 135–147.

Kordi, M., & Fotheringham, A. S. (2016). Spatially weighted interaction models—SWIM. *Annals of the American Association of Geographers*, *106*(5), 990–1012.

Lin, J. M., & Billa, L. (2021). Spatial prediction of flood-prone areas using geographically weighted regression. *Environmental Advances*, *6*, 100118.

Liu, N., Zou, B., Li, S., Zhang, H., & Qin, K. (2021). Prediction of PM2.5 concentrations at unsampled points using multiscale geographically and temporally weighted regression. *Environmental Pollution*, *284*, 117116.

Masarotto, G., & Varin, C. (2017). Gaussian copula regression in R. *Journal of Statistical Software*, *77*(8), 1–26.

Miller, J. A. (2012). Species distribution models: Spatial autocorrelation and non-stationarity. *Progress in Physical Geography: Earth and Environment*, *36*(5), 681–692.

Nakaya, T., Fotheringham, A. S., Brunsdon, C., & Charlton, M. E. (2005). Geographically weighted Poisson regression for disease association mapping. *Statistics in Medicine*, *24*(17), 2695–2717.

Nüst, D., Granell, C., Hofer, B., Konkol, M., Ostermann, F. O., Sileryte, R., & Cerutti, V. (2018). Reproducible research and GIScience: An evaluation using AGILE conference papers. *PeerJ*, *6*, e5072.

Openshaw, S. (1984). *The modifiable areal unit problem. Concepts and techniques in modern geography 38*. Norwich: GeoBooks.

Osborne, P. E., Foody, G. M., & Suárez-Seoane, S. (2007). Non-stationarity and local approaches to modelling the distributions of wildlife. *Diversity and Distributions*, *13*(3), 313–323.

Páez, A., Uchida, T., & Miyamoto, K. (2002). A general framework for estimation and inference of geographically weighted regression models: 1. Location-specific kernel bandwidths and a test for locational heterogeneity. *Environment and Planning A*, *34*(4), 733–754.

Sá, A. C. L., Pereira, J., Charlton, M. E., Mota, B., Barbosa, P. M., & Fotheringham, A. S. (2011). The pyrogeography of sub-Saharan Africa: A study of the spatial non-stationarity of fire—environment relationships using GWR. *Journal of Geographical Systems*, *13*(3), 227–248.

Sachdeva, M., Fotheringham, A. S., Li, Z. and Yu, H. (2023). On the local modeling of count data: Multiscale geographically weighted Poisson regression. *International Journal of Geographic Information Science*, at press.

Schaefer, F. K. (1953). Exceptionalism in geography: A methodological examination. *Annals of the Association of American Geographers*, *43*(3), 226–249.

Selby, B., & Kockelman, K. M. (2013). Spatial prediction of traffic levels in unmeasured locations: Applications of universal kriging and geographically weighted regression. *Journal of Transport Geography*, *29*, 24–32.

Simpson, E. H. (1951). The interpretation of interaction in contingency tables. *Journal of the Royal Statistical Society: Series B (Methodological)*, *13*(2), 238–241.

Wang, K., Zhang, C. R., Li, W. D., Lin, J., & Zhang, D. X. (2014). Mapping soil organic matter with limited sample data using geographically weighted regression. *Journal of Spatial Science*, *59*(1), 91–106.

Windle, M. J. S., Rose, G. A., Devillers, R., & Fortin, M.-J. (2010). Exploring spatial non-stationarity of fisheries survey data using geographically weighted regression (GWR): An example from the Northwest Atlantic. *ICES Journal of Marine Science*, *67*(1), 145–154.

Wu, C., Ren, F., Hu, W., & Du, Q. (2019). Multiscale geographically and temporally weighted regression: Exploring the spatiotemporal determinants of housing prices. *International Journal of Geographical Information Science*, *33*(3), 489–511.

Wu, S., Wang, Z., Du, Z., Huang, B., Zhang, F., & Liu, R. (2021). Geographically and temporally neural network weighted regression for modeling spatiotemporal non-stationary relationships. *International Journal of Geographical Information Science*, *35*(3), 582–608.

Ye, H., Huang, W., Huang, S., Huang, Y., Zhang, S., Dong, Y., & Chen, P. (2017). Effects of different sampling densities on geographically weighted regression kriging for predicting soil organic carbon. *Spatial Statistics*, *20*, 76–91.

Zhang, L., Cheng, J., Jin, C., & Zhou, H. (2019). A multiscale flow-focused geographically weighted regression modelling approach and its application for transport flows on expressways. *Applied Sciences*, *9*(21), 4673.

Zhang, L., Gove, J. H., & Heath, L. S. (2005). Spatial residual analysis of six modeling techniques. *Ecological Modelling*, *186*(2), 154–177.

Zhang, L., Ma, Z., & Guo, L. (2009). An evaluation of spatial autocorrelation and heterogeneity in the residuals of six regression models. *Forest Science*, *55*(6), 533–548.

Zhang, Z., Li, J., Fung, T., Yu, H., Mei, C., Leung, Y., & Zhou, Y. (2021). Multiscale geographically and temporally weighted regression with a unilateral temporal weighting scheme and its application in the analysis of spatiotemporal characteristics of house prices in Beijing. *International Journal of Geographical Information Science*, *35*(11), 2262–2286.

Index

G

H

I

K

L

M